Customer and Business Analytics

Applied Data Mining for Business Decision Making Using R

Chapman & Hall/CRC
The R Series

Series Editors

John M. Chambers
Department of Statistics
Stanford University
Stanford, California, USA

Torsten Hothorn
Institut für Statistik
Ludwig-Maximilians-Universität
München, Germany

Duncan Temple Lang
Department of Statistics
University of California, Davis
Davis, California, USA

Hadley Wickham
Department of Statistics
Rice University
Houston, Texas, USA

Aims and Scope

This book series reflects the recent rapid growth in the development and application of R, the programming language and software environment for statistical computing and graphics. R is now widely used in academic research, education, and industry. It is constantly growing, with new versions of the core software released regularly and more than 2,600 packages available. It is difficult for the documentation to keep pace with the expansion of the software, and this vital book series provides a forum for the publication of books covering many aspects of the development and application of R.

The scope of the series is wide, covering three main threads:
- Applications of R to specific disciplines such as biology, epidemiology, genetics, engineering, finance, and the social sciences.
- Using R for the study of topics of statistical methodology, such as linear and mixed modeling, time series, Bayesian methods, and missing data.
- The development of R, including programming, building packages, and graphics.

The books will appeal to programmers and developers of R software, as well as applied statisticians and data analysts in many fields. The books will feature detailed worked examples and R code fully integrated into the text, ensuring their usefulness to researchers, practitioners and students.

Published Titles

Customer and Business Analytics: Applied Data Mining for Business Decision Making Using R, *Daniel S. Putler and Robert E. Krider*

Event History Analysis with R, *Göran Broström*

Programming Graphical User Interfaces with R, *John Verzani and Michael Lawrence*

R Graphics, Second Edition, *Paul Murrell*

Statistical Computing in C++ and R, *Randall L. Eubank and Ana Kupresanin*

The R Series

Customer and Business Analytics

Applied Data Mining for Business Decision Making Using R

Daniel S. Putler

Robert E. Krider

CRC Press
Taylor & Francis Group
Boca Raton   London   New York

CRC Press is an imprint of the
Taylor & Francis Group an **informa** business

A CHAPMAN & HALL BOOK

CRC Press
Taylor & Francis Group
6000 Broken Sound Parkway NW, Suite 300
Boca Raton, FL 33487-2742

© 2012 by Taylor & Francis Group, LLC
CRC Press is an imprint of Taylor & Francis Group, an Informa business

No claim to original U.S. Government works

Printed in the United States of America on acid-free paper
Version Date: 20120327

International Standard Book Number: 978-1-4665-0396-0 (Paperback)

This book contains information obtained from authentic and highly regarded sources. Reasonable efforts have been made to publish reliable data and information, but the author and publisher cannot assume responsibility for the validity of all materials or the consequences of their use. The authors and publishers have attempted to trace the copyright holders of all material reproduced in this publication and apologize to copyright holders if permission to publish in this form has not been obtained. If any copyright material has not been acknowledged please write and let us know so we may rectify in any future reprint.

Library of Congress Cataloging-in-Publication Data

Putler, Daniel S.
   Customer and business analytics : applied data mining for business decision making using R / Daniel S. Putler, Robert E. Krider.
      p. cm. -- (Chapman & Hall/CRC the R series)
   Includes bibliographical references and index.
   ISBN 978-1-4665-0396-0 (pbk. : alk. paper)
   1. Database marketing--Software. 2. Data mining. 3. Decision making--Data processing. 4. R (Computer program language) 5. Database management. I. Krider, Robert E. II. Title.

HF5415.126.P88 2012
658.4'0302855133--dc23   2012008925

Visit the Taylor & Francis Web site at
http://www.taylorandfrancis.com

and the CRC Press Web site at
http://www.crcpress.com

To our parents:
Ray and Carol Putler
Evert and Inga Krider

Contents

II   Predictive Modeling Tools   31

3 Basic Tools for Understanding Data   33

4 Multiple Linear Regression   81

11 K-Centroids Partitioning Cluster Analysis 259

List of Figures

List of Tables

Preface

In writing this book we have three primary objectives. First, we want to provide the reader with an understanding of the types of business problems that advanced analytical tools can address and to provide some insight into the challenges that organizations face in taking profitable advantage of these tools.

Our second objective is to give the reader an intuitive understanding of how different data mining algorithms work. This discussion is largely non-mathematical in nature. However, in places where we think the mathematics is an important aid to intuitive understanding (such as is the case with logistic regression), we provide and explain the underlying mathematics. Given the proper motivation, we think that many readers will find the mathematics to be less intimidating than they might have first thought, and find it useful in making the tools much less of a "black box."

The book's final primary objective is to provide the reader with a readily available "hands-on" experience with data mining tools. When we first started teaching the courses this book is based on (in the late 1990s), there were not many books on business and customer analytics, and the books that were available did not take a hands-on approach. In fairness, given the license costs of user-friendly data mining tools at that time (and commercial software products up to the present day), writing such a book was simply not possible. We both are firm believers in the "learning by doing" principal, and this book reflects this. In addition to hands-on use of software, and the application of that software to data that address the types of problems real organizations face, we have also made an effort to inform the reader of the issues that are likely to creep up in applied data mining projects, and present the CRISP-DM process model as a practical framework for organizing these projects.

This book is intended for two different audiences, but who we think have similar needs. The most obvious is students (and their instructors) in MBA and advanced undergraduate courses in customer and business analytics and applied data mining. Perhaps less apparent are individuals in small- to medium-size organizations (both businesses and not-for-profits) who want to use data mining tools to go beyond database reporting and OLAP tools in order to improve the performance of their organizations. These individuals may have job titles related to marketing, business development, fund raising, or IT, but all see potential benefits in bringing improved analytics capabilities to their organizations. We have come in contact with many people who helped bring the use of analytics to their organizations. A common theme that emerged from our conversations with these individuals is that the first applications of customer and business analytics by an organization are typically skunkworks projects, with little or no budget, and carried out by an individual or a very small team of people using a learn-as-you-go approach. The high cost of easy-to-use commercial data mining tools (a project that requires multiple thousands of dollars per seat software licenses is no longer a skunkworks project) and a lack of appropriate training materials are often major impediments to these projects. Instead, many of these projects are based on experiments that push Excel beyond its useful limits. This book, and its accompanying R-based software (R Development Core Team, 2011), provides individuals in small and medium-sized organizations with the skills and tools needed to successfully, and less painfully, start to develop an advanced analytics capability within their organizations.

The genesis of this book was an applied MBA-level business data mining course given by Dan Putler at the University of British Columbia that was offered on an experimental basis in the spring term of the 1998–1999 academic year. One of the goals of the experimental course was to determine if the nature of the material would overwhelm MBA students. The course was project based (with the University's Development organization being the first client), and used commercial data mining software from a major vendor, along with the training materials developed by that vendor. The experiment was considered a success, so the following year the course became a regular course at UBC, and, partially based on Dan's original materials, Bob Krider developed a similar course at Simon Fraser University for both MBA and undergraduate business students.

We soon decided that the vendor's training materials did not fully meet the needs of the course, and we began to jointly develop a full set of our own tutorials for the vendor's software that better met the course's needs. While our custom tutorials were a major improvement, we soon felt the need to use tools based on R, the widely used open source and free statistical software. There were several reasons for this. First, the process of students moving out of computer labs and onto their own laptops to do computer-oriented coursework was well under way, and the ability of our students to install the commercial software on their own machines suffered from both licensing and practical limitations. Second, our experience was that students often questioned the value of the time spent learning expensive, specialized software tools as part of a class since many of them believed, correctly, that their future employers would not have licenses for the tools, and they themselves would not have the funds to procure the needed software. These concerns are greatly reduced through the use of mature, open-source tools, since students know the tools will be readily available for free in the future. Third, as we discuss above, we wanted a means by which to meet the needs and financial constraints of individuals in small and medium-size organizations who want to experiment with the use of analytics in their own organizations. Finally, we, like many other academic researchers, were using R to conduct our research (which is robust, powerful, and flexible), and knew it was only a matter of time before R would extensively be used in industry as well, a process that is now well on its way.

While we do our research using R in the "traditional way" (i.e., using the R console's command line interface to issue commands, run script files, and conduct exploratory analyses), a command line interface is a hard sell to most business school students and to individuals in organizations who are interested in learning about and experimenting with data mining tools. Fortunately, at the time we were thinking about moving to R for our courses, John Fox (2005) had recently released the R Commander package, which was intended to be a basic instructional graphical user interface (GUI) for R. This became the basis of the R-based software tools used in this book. Originally we developed a custom version of the R Commander that included functionality needed for data mining, and we contributed a number of functions back to the original R Commander package that were consistent with John's goal of creating a basic instructional GUI for statistics education. Since its introduction, the flexibility of the R Commander package has greatly increased, and it now has an excellent plug-in architecture that allows for very customized tool sets, such as the RcmdrPlugin.BCA package that contains the software tools used for this book.

In addition to John Fox, there are a number of other people we would like to thank. First we would like to thank multiple years of students at the University of British Columbia, Simon Fraser University, and City University of Hong Kong who used draft chapters of the book in courses taught by us and our Simon Fraser University colleague Jason Ho. The students pointed out areas where explanations needed to be clearer, where the tutorials were not exactly right, and a very long list of typographical errors. Their input over the years has been extremely important in shaping this book. Nicu Gandilathe (BCAA) and Matt Johnson (Intrawest) gave us valuable input about how to make the book and the software more useful to customer and business analytics practitioners. We have greatly benefited from conversations and advice given by our colleagues John Claxton (UBC), Maureen Fizzell (SFU), Andrew Gemino (SFU), Ward Hanson (Stanford), Kirthi Kalyanam (Santa Clara University), Geoff Poitras (SFU), Chuck Weinberg (UBC), and Judy Zaichowski (SFU) on both the content of the book and the process of getting a book published. Our editor at CRC Press, Randi Cohen, has been a real pleasure to work with, quickly addressing any questions we have had, and making every effort to help us when we needed help. We also want to thank Doug MacLachlan (University of Washington) for his review of draft versions of this manuscript; he has helped to keep us honest. Lastly, and perhaps most important, we want to thank both of our families, especially our wives, Liza Blaney and Clair Krider, for the patience and support they have shown us while writing this book, including Dan's dad, who kept the pressure on by frequently asking when the book would be finished.

Daniel S. Putler, Sunnyvale, CA, USA
Robert E. Krider, Burnaby, BC, Canada

Part I

Purpose and Process

Chapter 1

Database Marketing and Data Mining

As recently as the early 1970s, most organizations either had little information about their interactions with customers or little ability to access (short of physically examining the contents of paper file folders) and act upon what information they did have for marketing purposes. The intervening 40 years has seen an ongoing revolution in the information systems used by companies. The lowering of computing and data storage costs have been the driving force behind this, making it economically feasible for firms to implement transactional databases, data warehouses, customer relationship management systems, point of sales systems, and the other software and technology tools needed to gather and manage customer information. In addition, a large number of firms have created loyalty and other programs that their customers gladly opt into that, in turn, allows these firms to track the actions of individual customers in a way that would otherwise not be possible.

While falling computing costs and software advances allowed companies to develop increasingly sophisticated databases containing information about their interactions with their own customers, third-party data suppliers have taken advantage of the same information technology advances to collect additional information about those same customers, along with information on potential new customers, using data from credit reporting services, public records, the census, and other sources. As a result, companies now have the potential to prospect for new customers by finding individuals and organizations that are similar in important respects to their existing customers.

Realizing the potential of this newly available customer information has been a challenge to many organizations. While even small organizations now have the ability to develop extensive customer databases, up to now, only a fairly small number of comparatively large organizations have been able to take full advantage of the extensive information assets available to them. To do this, these firms have invested in analytical capabilities, particularly data mining, to develop managerially useful information and insights from the large amounts of raw data available.

The benefits of using these analytical tools are both practical/tactical and strategic in nature. From a practical/tactical perspective, the use of data mining tools can greatly reduce costs by better targeting existing customers, minimizing losses due to fraud, and more accurately qualifying potential new customers. In addition to lowering marketing costs, these tools can assist in both maintaining and increasing revenues through helping to obtain new customers, and in holding on (and selling more) to existing customers.

From a strategic point of view, organizations are increasingly viewing the development of the analytical capabilities needed to make the most of their data as a long-run competitive advantage. As Thomas Davenport (2006) writes in the *Harvard Business Review*:

> Most companies in most industries have excellent reasons to pursue strategies shaped by analytics. Virtually all the organizations we identified as aggressive analytics competitors are clear leaders in their fields, and they attribute much of their success to the masterful exploitation of data. Rising global competition intensifies the need for this sort of proficiency. Western companies unable to beat their Indian or Chinese competitors on product cost, for example, can seek the upper hand through optimized business processes.

The goal of this book is to provide you, the reader, with both a better understanding of what these analytical tools are and the ability to apply these tools to your own business, particularly as it relates to the marketing function of that business. To start this process, this chapter provides an overview of both database marketing and the data mining tools needed to implement effective database marketing programs.

1.1   Database Marketing

The fundamental requirement for any database marketing program is the development and maintenance of a customer database. In their book *The One to One Future*, Peppers and Rogers (1993) provide the following definition of a customer database:

> A **Customer Database** is an organized collection of comprehensive data about individual customers or prospects that is current, accessible, and actionable for such marketing purposes as lead generation, lead qualification, sale of a product or service, or maintenance of customer relationships.

In turn, Peppers and Rogers (1993) define database marketing in the following way:

Database Marketing is the process of building, maintaining, and using customer databases and other databases for the purposes of contacting and transacting.

1.1.1 Common Database Marketing Applications

The above definitions provide a useful starting point, but are a bit abstract. Looking at the most common types of database marketing applications should help make things clearer. Database marketing applications can be placed into three broad categories: (1) selling products and services to new customers; (2) selling additional products and services to existing customers; and (3) monitoring and maintaining existing customer relationships. The two most common types of applications designed to assist in the selling of products and services to new customers are "prospecting" for (i.e., finding) new customers, and qualifying (through activities such as credit scoring) those potential new customers once they have been found.

Database marketing applications designed to sell more to existing customers include cross-selling, up-selling, market basket analysis, and recommendation systems. Cross-selling involves targeting a current customer in order to sell a product or service to that customer that is different from the products or services that customer has previously purchased from the organization. An example of this is a telephone service provider who targets an offer for a DSL subscription package to a customer who currently only purchases residential land line phone service from that provider. In contrast, up-selling involves targeting an offer to an existing customer to upgrade the product or service he or she is currently purchasing from an organization. For instance, a life insurance company that targets one of its current term life insurance policy holders in an effort to move that customer to a whole life policy would be engaged in an up-selling activity.

Market basket analysis involves examining the composition of items in customers' "baskets" on single purchase occasions. Given its nature, market basket analysis is most applicable to retailers, particularly traditional brick and mortar retailers. The goal of the analysis is to find merchandising opportunities that could lead to additional product sales. In particular, a supermarket retailer may find that people who buy fresh fish on a purchase occasion are disproportionately likely to purchase white wine as well. As a result of this finding, the retailer might experiment with placing a display rack of white wine adjacent to the fresh fish counter to determine whether this co-location of products increases sales of white wine, fresh fish, or both.

Common applications designed to monitor and improve customer relationships include customer attrition (or "churn") analysis, customer segmentation, recommendation systems, and fraud detection. The goal of churn analysis is to find patterns in a current customer's purchase and/or complaint behavior that suggests that the customer is about to become an ex-customer. Knowing whether a profitable customer is at risk of leaving allows the organization to proactively communicate with the customer in order to present a promotional offer or address the customer's concerns in an effort to keep that customers business. Alternatively, a company may avoid taking actions that would encourage an unprofitable customer to remain with the firm. Grouping customers into segments based on their past purchase behavior allows the organization to develop customized promotions and communications for each segment, while recommendation systems, such as the one used by Amazon.com, group products based on which customers have bought them, and then makes recommendations based on the overlap of the buyers of two or more products. Fraud detection allows an organization to uncover customers who are engaged in fraudulent behavior against them. For instance, a consumer package goods company may use data on manufacturer's coupon redemptions on the part of different retail trade accounts in order to develop a model that would flag a particular retail account as being in need of further investigation to determine whether that retailer is fraudulently redeeming bogus coupons that were not actually redeemed by final consumers.

Two Examples

To get a sense of how organizations use database marketing in practice, we examine two different database marketing efforts. The first is an application designed to prospect for new customers, while the second deals with two related projects designed to reduce customer churn. One thing that is common to both these applications is that there are substantial savings in marketing costs (that more than cover the analysis costs) from not conducting blanket promotions.

Keystone Financial

In his article "Digging up Dollars with Data Mining—An Executive's Guide," Tim Graettinger (Graettinger, 1999; Kelly, 2003) describes a database marketing project undertaken by Pennsylvania-based Keystone Financial Bank, a regional bank. Keystone developed a promotional product called LoanCheck with the intention of using it to expand its customer base (a prospecting application). LoanCheck consisted of a $5,000 "check" that could be "cashed" by the recipient at any Keystone Financial Bank branch to initiate a $5,000 loan. To determine which potential new customers Keystone should target with this product, Keystone mailed a LoanCheck offer to its existing customers. Information on which of its existing customers took advantage of the LoanCheck offer was appended to Keystone's customer database. The customer database was then used to determine the characteristics of customers most likely to respond favorably to the LoanCheck offer using data mining methods, resulting in the creation of a model that predicted the relative likelihood that a customer would respond favorably to the LoanCheck offer. Keystone then applied this model to a database of 400,000 potential new customers it obtained from a credit reporting agency, and then mailed the LoanCheck offer to the set of individuals in that database the model predicted would be most likely to respond favorably to the LoanCheck offer. This database marketing project resulted in Keystone obtaining 12,000 new customers, and earning $1.6M in new revenues.

Verizon Wireless

At the 2003 Teradata Partners User Group Conference and Expo, Ksenija Krunic, head of data mining at Verizon Wireless (a major U.S. mobile phone service provider), described how her company used two related database marketing projects to decrease Verizon Wireless's churn rate for individual customers by one-quarter compared to what it had been (Das, 2003). Specifically, in the first project, Verizon used its customer databases in order to develop a model to predict which of its customers were most likely to defect to another provider at the expiration of their current contract based on the current plan a customer had, a customer's historical calling patterns, and the number and type of service requests made by a customer. The second project involved using the model developed in the first project to create samples of customers likely to leave Verizon at the end of their current contract, and then offer each of these samples a different experimental new plan offer, tracking which customers in each segment accepted the offer (thereby resulting in a contract renewal with Verizon). The data generated from these experimental samples (which consisted of whether a customer took the service and the terms of the offered plan) were combined with the customer calling pattern and service request data to create a second set of models which, together, allow Verizon to determine the best new plan offer to make to a customer who is likely to leave Verizon at the end of his or her contract, before the current contract expires, including not making an offer at all. Using these models, Verizon Wireless was able to decrease its attrition rate from 2 percent per month to 1.5 percent per month (a reduction of 25 percent from the original attrition rate). Given that the cost of acquiring a customer in the mobile phone industry is estimated to be between $320 and $360, the drop in the attrition rate has had a huge impact on Verizon Wireless's bottom line. Verizon has 34.6 million subscribers, so the value of the reduction in churn is roughly $700M per year. In addition, since the promotional mailings are now highly targeted, the company's direct mail budget for "churner mailings" fell 60 percent from what it was prior to the completion of these two related database marketing projects.

1.1.2 Obstacles to Implementing a Database Marketing Program

As the above two examples indicate, the potential rewards from implementing database marketing programs can be enormous. Unfortunately, there are a number of obstacles that can make implementing these programs difficult. First, the data issues can be complex. Specifically, IT systems and tools (such as data warehouses and customer relationship management systems) need to be in place to collect the needed data, clean the data, and integrate data that can come from a large number of different computer systems, databases, and Excel spreadsheets. Second, the data mining tools themselves can be complex since they are based on a combination of advanced statistical and machine learning tools. Finally, the available talent that can be hired who "can do it all" in terms of understanding both the analytical tools and the business problems is scarce. As Davenport (2006) writes: "Analytical talent may be to the early 2000s what programming talent was to the late 1990s. Unfortunately, the U.S. and European labor markets aren't exactly teaming with analytically sophisticated job candidates."

While these three obstacles are not insurmountable, it can take a considerable amount of time and effort to overcome them. The experience of Barclays Bank, as described by Davenport (2006), illustrates this point:

> The UK Consumer Cards and Loans business within Barclays bank, for example, spent five years executing its plan to apply analytics to the marketing of credit cards and other financial products. The company had to make process changes in virtually every aspect of its consumer business: underwriting risk, setting credit limits, servicing accounts, controlling fraud, cross selling, and so on. On the technical side, it had to integrate data on 10 million Barclaycard customers, improve the quality of the data, and build systems to step up data collection and analysis. In addition, the company embarked on a long series of small tests to begin learning how to attract and retain the best customers at the lowest price. And it had to hire new people with top-drawer quantitative skills.

Despite the obstacles, the use of data mining–based database marketing continues to grow. Evidence of this is that the dollar sales of the software tools needed to implement this type of analysis grew 11.5 percent in 2005 over 2004 levels, and industry forecasts made by IDC (Vesset and McDonough, 2006) indicate that this rate of growth will be maintained for the foreseeable future.

1.1.3 Who Stands to Benefit the Most from the Use of Database Marketing?

While most organizations can obtain some benefit from the use of database marketing tools, some will receive substantially greater benefits than others. Three factors are particularly important in driving the returns to database marketing programs: (1) the organization has a large number of customers; (2) customer transaction data can be obtained either as a byproduct of normal operations or through the use of a device, such as a customer loyalty program, by the organization; and (3) the acquisition and/or loss of a customer is expensive to the organization.

Given the nature of these three factors, it is unsurprising that certain industries have emerged as leaders in implementing database marketing programs. These leading industries include (1) telecommunications; (2) banking, insurance, and financial service providers; (3) catalog and online retailers; (4) traditional retailers; (5) airlines, hotel chains, and other travel industry players; and (6) charities, educational institutions, and other not-for-profits.

1.2 Data Mining

As the examples in the previous section indicate, the underlying technology driving database marketing efforts is data mining. In this section we provide an overview of data mining by first providing two definitions of what it is, and then briefly describing commonly used data mining methods. Our descriptions of methods in this chapter are brief (at times almost non-existent) since the balance of the remainder of this book covers these methods in much greater detail.

1.2.1 Two Definitions of Data Mining

Data mining really has two different intellectual roots, statistics and the database and machine leaning fields of computer science. Because of this twin heritage, a large number of different definitions of data mining have been put forward. Probably the most widely used definition of data mining comes from The Gartner Group (Krivda, 1996):

> **Data mining** is the process of discovering meaningful new correlations, patterns, and trends by sifting through large amounts of

data stored in repositories and by using pattern recognition technologies as well as statistical and mathematical techniques.

This definition of data mining flows more from the database and machine learning tradition. In this tradition, data mining is also referred to as "knowledge discovery in databases" or KDD. A common theme in this tradition is that the application of data mining methods to data will reveal new, heretofore unknown patterns that can then be constructively taken advantage of. This world view is in marked contrast to the one of traditional statistics, where patterns are hypothesized to exist a priori, and then statistical methods are used to test whether the hypothesized patterns are supported by the data.

To reveal our bias, we lean toward the statistics world view. The machine learning world view strikes us as being a bit too "auto-magical" for our tastes. Moreover, given our econometrics-oriented training and backgrounds, we are concerned about both spurious correlation and attempting to gain additional insight by understanding the drivers of customer behavior. As a result, we place a lot of emphasis on modeling behavior as a means of predicting it. Given this orientation, the definition of data mining we use is:

> **Data mining** is the process of using software tools and *models* to summarize large amounts of data in a way that supports decision-making.

The critical difference in our definition is its focus on models and modeling. We view modeling as the human process of simplifying a complex real world situation by abstracting essential elements. Properly done, modeling improves our understanding, our ability to communicate, and our decision-making.

1.2.2 Classes of Data Mining Methods

Ultimately, data mining uses a set of methods that originated in either statistics or machine learning to summarize the available data. These different methods fall into two broad classes, grouping methods and predictive modeling methods. Within each of these two classes fall literally hundreds of different specific methods (also known as algorithms). In this section we will only mention the most commonly used methods for each of the two classes. We will present these methods in more detail later in the book.

1.2.2.1 Grouping Methods

Grouping methods used in database marketing can be categorized as falling into two distinct types: methods used to group products and services, and methods used to group customers. The most commonly used method to group products and services is known as *association rules*. Association rules come from machine learning, and examine the co-occurrence of different objects (say the purchase of fresh fish and white wine by customers on the same shopping occasion) and then form a set of "rules" that describe the nature of the most common co-occurrence relationships among objects in a database.

Two methods are commonly used to group customers. The most widely applied is cluster analysis, which is a term used to describe a set of related methods that were developed in statistics (some of the methods date to the 1930s). The most common method of cluster analysis used in data mining is known as *K-Means*. K-Means is one of several "partitioning methods" for cluster analysis that have been developed. K-Means is called a partitioning method since it finds the "best" (using a Euclidean distance-based measure) division of the data into K partitions, where K is the number of partitions specified by the analyst. The other commonly used methods of cluster analysis are known as hierarchical agglomerative methods (Wards method, average linkage, and complete linkage are the most commonly used hierarchical agglomerative methods). Hierarchical agglomerative methods are not typically used in data mining because they do not scale to the number of records often encountered in database marketing applications. However, these methods are well suited to the number of records typically used in sample survey–based marketing research applications.

The second method commonly used to group customers is known as *self-organizing maps* (also called Kohonen maps, after the inventor of the method, Finnish computer scientist Teuvo Kohonen). Euclidean distance is also used as the basis of grouping records in this method. However, how these distances are used is very different across the two methods. K-Means attempts to minimize the sum of the squared Euclidean distances for members within a group, while self-organizing maps use the distances as part of a neural network algorithm. One drawback to both of these methods is that the variables used to group records must be continuous, so categorical variables (such as zip or postal code) cannot be used to group customers. However, there are other clustering methods (such as ROCK clustering; Guha et al. (2000)) that can cluster a set of categorical variables.

1.2.2.2   Predictive Modeling Methods

Three types of methods are commonly used to construct predictive models in data mining: (1) linear and logistic regression; (2) decision trees; and (3) artificial neural networks. Consistent with the class name, the goal of all three methods is to predict a variable of interest. The variable can be either continuous (e.g., total sales of a particular product in the next quarter) or categorical (e.g., whether a customer will respond favorably to a particular direct mail offer) in nature. In the case of a continuous variable, what is predicted is the expected value of that variable (e.g., expected total sales of the product in the next quarter), while in the case of a categorical variable, what is predicted is the probability that the variable will fall into each of the possible categories (e.g., the probability a customer will respond favorably to the direct mail offer).

Linear and logistic regression are two of the most important tools of traditional statistical inference. Both methods use a weighted sum of an analyst-specified set of predictor variables (known as a "linear predictor") to come up with a predicted value. Where the two methods differ is in how this linear predictor is transformed in order to make a prediction. In the case of linear regression, the linear predictor constitutes the prediction, while in logistic regression the linear predictor is transformed in a way such that the predicted probability for each possible category of the categorical variables of interest falls between zero and one, and the sum of the probabilities across the different categories equals one. Both a plus and minus of linear and logistic regression is that the analyst plays a central role in creating a model. The plus to this is that the implied customer behavior underlying a model can be more easily seen, so it is easier for managers to interpret, critique, and learn from that model. The minus is that the quality of a model is closely tied to the skill level of the analyst who created it.

Decision tree methods have origins in both statistics and machine learning. While a number of different algorithms have been proposed (and are commonly used) to create a decision trees, all methods create a set of "if-then" rules leading to a set of final values for the variable being predicted. These final values can be either probabilities for a categorical variable (in which case the tree that is created is known as a "classification tree") or quantities for a continuous variable (where the resulting tree is called a "regression tree"). To give a better sense of what a decision tree looks like, Figure 1.1 shows a hypothetical classification tree of a churn analysis for a mobile telephone service provider.

The example classification tree starts at its "root" with a split on whether the customer had more or less than 100 calling minutes on average each month. If the answer to this question is no, we move to the next "node" where the split is determined based on whether the customer has a subscription to the "Basic" plan. If the answer to this question is yes, then the probability the customer will stay is 85 percent (or a 15 percent probability of leaving), while the probability of a customer staying with the company is only 10 percent if that customer had less than 100 calling minutes per month on average *and* the customer had a contract for something other than the basic plan. If the

Figure 1.1: An Example Classification Tree

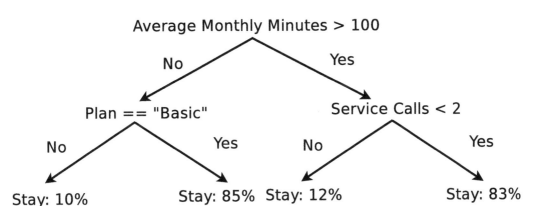

customer had more than 100 calling minutes per month on average, then the second node in the tree is the number of service calls the customer made. Each element at the bottom level of the tree that indicates the probability of staying or leaving the service provider is called a "leaf."

One advantage of decision trees is that most people find their "if-then" structure to be both easy to understand and to act upon. Another advantage is that less skilled analysts will get results similar to those of more skilled analysts since all the variables in the database can be used as predictors in a decision tree (the decision tree algorithm will determine which to include in the tree), and the algorithm automatically "transforms" the relevant variables via the splitting rules. However, decision trees also have a number of disadvantages as well, which we explore later.

An *artificial neural network* is a predictive modeling method developed in machine learning that is based on a simplified version of the brain's neurological structures. Figure 1.2 provides an illustration of a simple neural network. However, explaining even this simple example is fairly involved, so we will refrain from doing so now. The three important things to know at this point are that: (1) neural network models, like decision trees model, are less dependent on the skill of the analyst in developing a good model relative to linear and logistic regression; (2) neural network models are very flexible in terms of the shapes of relationships they can mimic, but this turns out to be something of a mixed blessing; and (3) neural network models are very hard to interpret in a managerially meaningful way, so they amount to "black boxes" that can predict well but provide no insights into underlying customer behavior.

Figure 1.2: An Example Neural Network

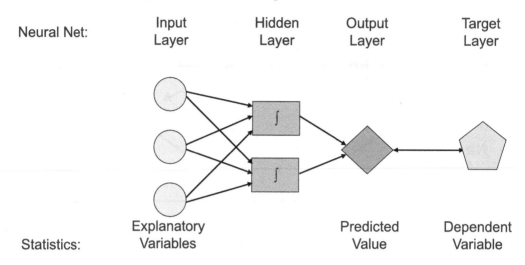

1.3   Linking Methods to Marketing Applications

It will probably come as no surprise that there is a strong relationship between the type of database marketing application being undertaken and the class of the data mining method that should be used for that application. Table 1.1 provides a table that relates each common type of database marketing application to the appropriate class of data mining method to use for that application. An examination of the table reveals that the only application type where it could make sense to use both grouping and predictive modeling methods is fraud detection, although, for this application, we expect that predictive modeling methods will be a superior choice to grouping methods. However, it may be possible to group customers in such a way that one of the groups formed has a higher incidence of fraudulent behavior, making grouping methods potentially useful.

Now that you have a better sense of what both database marketing is, and how data mining tools are used in database marketing programs, you are ready to learn more details about the process of implementing a data mining project as part of a database marketing program.

Table 1.1: Linking Marketing Applications with Data Mining Methods

Marketing Application	Data Mining Method Class	
	Grouping	Predictive Modeling
Prospecting		√
Prospect qualifying		√
Cross-selling		√
Up-selling		√
Market basket analysis	√	
Recommendation systems	√	
Attrition and churn		√
Fraud detection	√	√
Customer segmentation	√	

Chapter 2

A Process Model for Data Mining—CRISP-DM

In this chapter we examine a process model for data mining called CRISP-DM. CRISP-DM is an acronym for the CRoss-Industry Standard Process for Data Mining. CRISP-DM has become something of a de facto standard for organizing and conducting data mining projects. The document that describes CRISP-DM in detail is *CRISP-DM 1.0: Step-By-Step Data Mining Guide* (Chapman et al., 2000), which is published by the CRISP-DM consortium, and can be freely downloaded from the CRISP-DM Consortium's web site (www.crisp-dm.org or from this book's web site, www.customeranalyticsbook.com). Our goal is a condensed, somewhat paraphrased, overview of the CRISP-DM 1.0 model. Before doing this, however, we provide some historical context on data mining process models in general and CRISP-DM in particular.

2.1 History and Background

Business-oriented data mining only started to become something of an organized field in the early to mid-1990s. Very quickly, organizations developing internal database marketing and data mining capabilities, data mining consultants, and software vendors selling data mining tools came to realize there was a need to systematically organize the process of data mining. To this end, Barry and Linoff (1997) in their influential book *Data Mining Techniques for Marketing, Sales and Customer Relationship Management* presented a high-level model of the data mining process that they dubbed "the virtuous cycle of data mining." At roughly the same time, software vendors presented process guides for data mining that covered steps from the point at which data were loaded into data mining software tools through to the early stages of the deployment of a data mining–based solution. The best known of these process guides is SAS's SEMMA process (SEMMA is an acronym that stands for sample, explore, modify, model, and assess).

The broadest-based effort to develop a process model for data mining was started in 1996 through a joint effort of Daimler-Benz, ISL (which was later acquired by SPSS, which was, in turn, was recently acquired by IBM), and NCR. Daimler-Benz was an early adopter of data mining methods to address business problems; ISL developed the first data mining software workbench (Clementine, first marketed in 1994); and NCR's interest was based on its desire to add value to its Teradata data warehouse products, and it had entered the data mining consulting field in an effort to help accomplish this. The goal of this initial group was to help new adopters of data mining methods to avoid a long period of trial-and-error learning in implementing these methods to solve business problems. The group believed that the best way to do this was to develop a standard process model that would be non-proprietary, freely available, and data mining software tool neutral.

In 1997, partially supported with funding from the European Commission, the group (along with OHRA Verzekeringen en Bank Groep, a Dutch banking firm) founded a consortium under the CRISP-DM banner. An important part of the consortium's efforts was the creation of a special interest group (or SIG) that allowed for a broader set of individuals and organizations to play a role in the development of the CRISP-DM process model. Drafts of the CRISP-DM model were available as early as 1999, and the consortium released a final version (version 1.0) of the CRISP-DM model in 2000. At the time of this writing (September 2009), efforts are under way to develop version 2 of CRISP-DM.

The major competitor of the CRISP-DM process model is a fusing of Barry and Linoff's (Barry and Linoff, 1997) virtuous cycle of data mining with SAS's SEMMA approach for conducting the actual data analysis. The fusion of the virtuous cycle and SEMMA (which is now much more explicitly done as a result of Barry and Linoff's close ties with SAS) results in a process model that is fairly similar to CRISP-DM. However, CRISP-DM goes into aspects of conducting a data mining project beyond the process model itself. Put another way, the virtuous cycle model fused with SEMMA gives you the guidance you need to conduct a data mining project using best practices, while CRISP-DM does this, plus gives the user a head start in laying out the tasks and milestones in a Gantt chart for a particular project.

2.2 The Basic Structure of CRISP-DM

CRISP-DM really provides three different things: (1) a step-by-step "blueprint" for conducting a data mining project (the tasks of a Gantt chart); (2) a specified set of "deliverables" for each phase of a project (the milestones of a Gantt chart); and (3) a set of documentation standards in terms of what information should be included in each report that is in the set of project deliverables. The documentation standards are useful in that they provide the information needed to replicate a project if need be (which can be critical in certain industries), and to provide the basis of learning for future data mining projects. In this chapter we will only look at the process model, but the reader is encouraged to obtain the complete *CRISP-DM 1.0 Users Guide* at www.crisp-dm.org or www.customeranalyticsbook.com.

2.2.1 CRISP-DM Phases

The CRISP-DM process model is organized into a set of six phases. The phases fit into a loose natural order, but the order is not strict since in real-world projects it is understood that as additional information is uncovered, leading to new understandings, there will often be a need to adjust earlier phases in the project. The six phases are: (1) Business understanding; (2) Data understanding; (3) Data preparation; (4) Modeling; (5) Evaluation; and (6) Deployment. Figure 2.1 provides a graphical representation of the relationships between the phases of the CRISP-DM process model.

In addition to showing the general flow of the six phases, the figure also illustrates the most common types of interplay that occurs between phases. Specifically, a project will start by attempting to understand the business and then move to understanding the data available to address the identified business opportunity or problem. However, by examining this data, additional issues and questions about the business problem or how the business operates may emerge, which leads to possible changes in understanding about the business. Typically, there is also a great deal of two-way interplay between the data preparation and the modeling phases of a project, either to prepare the data in a way so that it suitable for use with a different modeling method than was originally planned, or to transform existing variables and/or derive new variables from existing variables that initial modeling suggests may be useful in predicting the behavior under investigation. Finally, it is not uncommon for a data mining project to provide new information and understanding, or new questions, about the business. This new understanding or the new questions

Figure 2.1: Phases of the CRISP-DM Process Model

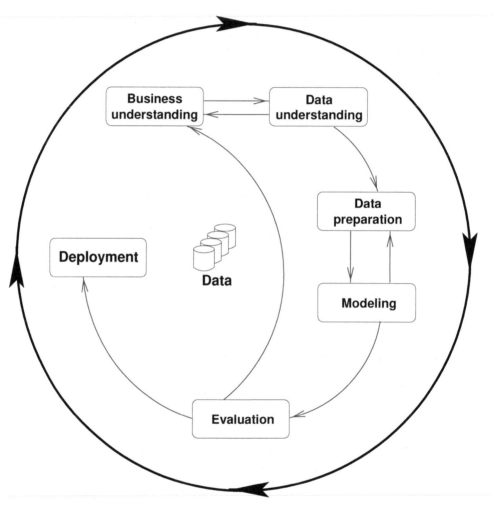

raised about the business may result in a rethinking about the business either for this project, or, more likely, may suggest other business opportunities or problems that management may want to address.

2.2.2   The Process Model within a Phase

Within each phase, the structure of the CRISP-DM process model is hierarchical in nature. At the top level of the hierarchy is the phase itself (e.g., business understanding). Underneath the phase will be several *generic tasks* (the number of generic tasks differs across phases). As the name suggests, generic tasks are high-level descriptions of activities that need to be carried out for nearly all projects. Each of these generic tasks is then made project specific through the specification of one or more *specialized tasks*, which fall under the generic task at the third level of the hierarchy. The specialized tasks should describe the specific activities that need to be undertaken to accomplish a generic task. For example, one of the generic tasks under the business understanding phase is to determine business objectives. Specialized tasks to address this generic task might include: (1) Review the annual marketing plan for product X for each of the past three years; (2) Schedule and conduct an interview with Amy Ng, Executive Vice President of Marketing; and (3) Schedule and conduct an interview with Gordon Stewart, product manager for product X. At the lowest level of the hierarchy are the *process instances* that specify the deliverable(s) associated with each task. In the example given above, process instances would include items such as a transcript and summary for each of the two interviews, a summary of product X's performance and historical objectives based on the review of product X's past marketing plans, and a listing of the current objectives for product X to address the generic task.

2.2.3   The CRISP-DM Phases in More Detail

In this section we describe in greater detail the six phases of the CRISP-DM process model and give a list of generic tasks for each phase. While additional detail is provided, the details given are far from complete. As stated earlier, we encourage you to obtain a complete copy of the *CRISP-DM 1.0 Users Guide*.

2.2.3.1   Business Understanding

The purpose of the business understanding phase is to develop a set of project objectives and requirements from a business perspective, and then converting those objectives and requirements into a definition of the data mining problem to be addressed, and creating a preliminary plan designed to address that problem.

The generic tasks of this phase are: (1) determine business objectives; (2) assess the situation; (3) determine data mining goals; and (4) produce a project plan. The point of the first generic task is to determine what the organization really wants to accomplish. Often there will be a number of competing objectives and constraints that will need to be uncovered and then assessed in order determine how to make appropriate trade-offs. Not doing this may lead you to provide the right answers to the wrong questions.

The *situation assessment task* requires additional fact-finding to determine the resources, constraints, assumptions, and other factors that potentially influence both the data mining goal and the project plan. Determining the data mining goal involves translating the business objectives into a set of data mining project goals in technical terms. For instance, the business goal of Verizon Wireless in the example given in the last chapter was to increase the renewal rate (reduce churn) of customers whose contracts were expiring, which leads to two related data mining questions. First, which customers were most likely to leave Verizon at the end of their current contracts? Second, how likely was a customer with a particular profile to accept an offer for a particular new plan that Verizon could offer in order to keep that customer's business?

The *project plan*, which is really the end result of this phase, should be comprehensive and detailed. In particular, the plan should specify the anticipated steps, in order, along with potential dependencies. The plan should also specify all inputs and outputs. The creation of this plan is likely to benefit from the use of formal project planning tools such as Gantt charts.

2.2.3.2 Data Understanding

The data understanding phase begins with determining what data are currently available to the organization, whether permission can be granted to use this data for data mining, if there are any restrictions on the use of the data (such as privacy concerns), and what data are relevant to addressing the data mining problem. In addition, there needs to be a determination of whether there are any applicable third-party data available (such as customer credit history data), and the details (such as cost, data structure, data format, etc.) concerning applicable third-party data.

The next part of this phase involves gathering the appropriate data, and then exploring the data to get an understanding of it, identifying any quality problems, and using simple descriptive statistics and visual displays in order to develop an initial understanding that will be of use in the modeling phase.

There are four generic tasks associated with the data understanding phase: (1) collect initial data; (2) describe data; (3) explore data; and (4) verify data quality. The *collect initial data task* deals with the issues surrounding locating, assessing, and obtaining the needed data (from both internal and third-party sources) discussed above. The *describe data task* involves examining the "surface" properties of the acquired data. The surface properties of the data include such things as the format of the data (e.g., a relational database table versus a comma separated value text file); the amount of data (in terms of the number of records and variables); and the names, data types (e.g., categorical or numeric), coding schemes, and definitions of the variables in the data. A key objective of this task is to determine whether the collected data will answer the identified data mining question(s).

The *explore data task* involves the use of frequency distributions, cross-tabulations, means, correlations, and other simple descriptive statistics of the variables of interest, along with a number of other variables initially thought to influence the variables of interest. In addition, graphical tools such as histograms, scatter plots, and other simple plots of the data are frequently useful at this stage. These analyses may help refine the data description, lead to a better understanding of potential data quality issues, and help gain a basic understanding of the nature of relationships between different variables that will be of use in the modeling phase.

The *verify data quality task* addresses the following important questions. Are the data complete (i.e., do they cover all the relevant cases we hope to examine)? Do all the variables seem to be correct (e.g., are there variables that should be all numbers that contain some character entries; is the earliest date of a transaction dated November 17, 1999, when transaction data should go back further than this)? Are there missing values in the data? If yes, how common are they, and why do they occur?

2.2.3.3   Data Preparation

In the data preparation phase the final dataset to be used in model building is constructed from the available raw data. The preparation may include selecting records to use in the analysis; creating "clean" samples of records to use in the modeling process; selecting the variables to use in the analysis; and creating new variables by transforming some of the variables in the raw data or deriving them based on two or more variables in the raw data. As indicated earlier in the chapter, it is not unusual to move back and forth between this phase and the modeling phase as the project progresses.

There are five generic tasks associated with the data preparation phase: (1) select data; (2) clean data; (3) construct data; (4) integrate data; and (5) format data. One of the most important steps in conducting a data mining project is *selecting the data* to use for the actual analysis. Data selection relates to both what variables to have available in the data set to be used for the actual data mining, as well as the nature of the data records to be used in the analysis. A number of issues come up with respect to the nature of the data records to include in the analysis. Of particular concern for predictive modeling applications is that often a positive response to a promotional offer (the behavior of interest) can represent a very small percentage of the customer base (response rates of around two percent for untargeted promotional offers are not uncommon). In these cases, separating signal from noise within the data can be very difficult. As a result, it is not uncommon to select records for analysis such that the positive responders to the variable of interest are over-sampled relative to the negative responders for the variable of interest. Moreover, some data mining methods (decision trees in particular) perform much better if there are roughly equal proportions of positive and negative responders in the dataset used in the data mining analysis. Thus, it is often the case that the dataset used in the data mining analysis consists of a stratified random sample of the original database in which there are equal numbers of positive and negative responders to the variable of interest, even though the positive responders represent only 2 percent of the original database. However, since we know the relative probabilities of including positive and negative responders, we can project the results based on the analysis dataset back to the overall customer database. Another question that needs to be addressed is how to handle records with missing data for variables that are thought likely to be important for the analysis. One solution is to simply omit records with important missing information. However, this may (or may not) lead to biased results.

The main element of the *data cleaning* task involves how to deal with data with missing values. Specifically, should missing values for a variable be replaced with a default value (say of zero or the mean value of the non-missing values for that variable), and if a default value is used, should another variable be created to indicate what records actually have missing values for that variable? Unfortunately, the answer to this question is "it depends." It turns out that at times non-responses to questions contain information in their own right. For instance, a credit card company in qualifying prospects may find that individuals who did not complete certain questions on their applications have a higher probability of defaulting on a credit card if it is issued compared to individuals who answered those questions.

Other variables with values that are likely to be incorrect must also be dealt with. To illustrate this point, one of the authors was once involved in a project to assist a charitable organization with their fundraising. An issue that became quickly apparent was that a number of variables containing date information had an extremely high number of records where the value of the date was November 17, 1999. Ultimately, it was determined that these values were erroneous, and were caused by a botched Y2K conversion of the charity's donor database. The problem then became one of how to deal with the erroneous date data. Fortunately, it turned out that other tables in the database had what was thought to be duplicate information, but in fact had the correct values for the records with corrupted date information, allowing for the erroneous values to be replaced with correct values. If this "duplicate" data was not available, then an important question that would need to have been addressed was whether only data with correct date information should be used in the analysis, thereby creating a "clean" subset of the data, but at the cost of discarding some of the records in the database.

The *construct data task* typically involves creating new variables through transforming a single variable (e.g., taking a natural logarithm of one of the variables in the original database) or creating derived variables from other variables in the database (e.g., dividing the total number of transactions a customer has made with our company by the number of months since that customer's first transaction to create an average transactions per month purchase frequency variable). The other thing this task may involve is creating completely new records. For example, there may be a need to create records for customers who have made no purchases in the past year if we are working with the past year's customer transactions database. By implicitly ignoring customers with no transactions, we may be overlooking important information.

The *integrate data task* involves merging different data tables together (say the transaction history of a customer, that is contained in the transactions database table, with the customer's personal information, contained in the customer information table) in order to create a single dataset that can be used by the data mining tools.

The *format data task* primarily refers to potential minor changes in the structure of variables or the order of variables in a database so that they match what is expected by a particular data mining method. Alternatively, it may involve changing the order of records so that the order is close to random so that certain data mining methods work properly. This issue tends to become most relevant if a stratified sample is created in order to oversample positive responses to the variable of interest, since the stratification may well result in all the records for the positive responders coming first in the new dataset, while the records for all the negative responders follow.

2.2.3.4 Modeling

As its name suggests, in the modeling phase the actual models are constructed and assessed. The generic tasks associated with this task are: (1) select modeling technique(s); (2) generate a test design; (3) build model; and (4) assess model.

As Table 1.1 presented at the end of Chapter 1 indicates, the selection of an appropriate modeling method(s) is dependent on the nature of the database marketing application. However, for most applications, there is more than one appropriate method. In the *select modeling technique(s)* task a decision is made as to which of the possible methods that can be used should be used. The decision could be made to use all applicable tools, and then select the model that is "best" among the set of possible models as part of the assess model task, which, in turn, relies on the testing procedures developed in the generate a test design task.

The *generate a test design task* needs to be done prior to building any models. The main purpose of the testing environment is to assess the quality and validity of different models. This is typically accomplished by taking the dataset created in the data preparation phase and dividing it into two or three different samples. The first of these samples is known as the estimation sample (it is also called the training sample). The purpose of this sample is to actually build the model(s). The second sample is called the validation or test sample, and its purpose is to examine the accuracy and validity of a particular model, and to provide a basis for comparing the accuracy of different models. Based on the validation sample, a "best" model can be selected. However, to get an unbiased estimate of the likely impact (in terms of sales and profits) of the use of this best model, a third sample is needed to make this assessment (which actually occurs in the evaluation phase of the process). This sample is known either as the holdout or validation sample.

The *build model task* is where the previously selected data mining methods are applied to the dataset. You will see exactly how to do this in later chapters.

In the *assess model task*, the focus is on assessing a model on its technical merits as opposed to its business merits. The main concern at this point is the accuracy and generality of the model. An assessment with respect to the business problem being addressed is done in the evaluation phase. The assessment of a model can result in the conclusion that the model can be improved upon, and also suggest ways of making an improvement, resulting in a new build model task.

2.2.3.5 Evaluation

At this point a model (or several) has been created that possesses a reasonable level of accuracy. Before deploying a model, an assessment of its likely impact needs to be made. This can partially be accomplished for predictive models by using the holdout sample to develop an estimate of the returns from using the model. However, this alone is not sufficient. Another critical factor that needs to be addressed is whether there is some important business issue that has not yet been sufficiently considered. In addition, the potential implications of repeated use of the model must be thought through to determine if there are any potential negative side effects associated with the use of the model.

The generic tasks of this phase are: (1) evaluate results; (2) review process; and (3) determine next steps. The activities associated with *evaluating the results* are discussed in the prior paragraph, while the *review process task* is really a quality assurance assessment, which addresses concerns such as: Was the model correctly built? Were the variables defined in a way that is consistent with the variables available in the actual customer database? Has a variable that is "trivially related" to the target variable (i.e., is a perfect predictor of the target since its value is based on the value of the target variable) been included in the model? Will the variables used in this analysis be available for future analyses? In the *determine next steps task*, the project team needs to decide whether to finish the project and move on to deployment (if appropriate) or whether to revisit certain phases of the project in an attempt to improve upon them.

2.2.3.6 Deployment

The nature of the deployment phase will vary with the nature of the project. Certain customer segmentation studies are done in an effort to gain a better understanding of an organization's customers, so deployment involves effectively communicating the knowledge gained from the project to appropriate people within the organization. In other cases, such as nearly all applications involving predictive modeling, there is a need to incorporate the estimated models into the organization's business decision processes, such as determining which customers should be targeted with a particular offer. In order to successfully deploy a data mining–based solution, four generic tasks may (depending on the type of project) need to be undertaken: (1) plan deployment; (2) plan monitoring and maintenance; (3) produce a final report; and (4) review project.

If relevant, a *deployment plan* needs to determine how best to incorporate any models developed into the relevant business processes of an organization. While this may seem straightforward, we have been involved in several projects where this proved to be a real stumbling block. It can be a real problem when consultants develop the data mining models, and there is insufficient buy-in on the part of personnel (particularly information systems personnel) in the client organization. Thus, an assessment of possible deployment pitfalls must be made. A related issue, but one relevant in almost all data mining applications, is determining who within the organization needs to be informed of the project's results, and how best to propagate this information.

Prior to deploying a data mining solution, *plans for monitoring and maintaining* that solution must be made. To do this, an assessment of what changes could occur in the future which would trigger an examination of the deployed model (the entrance of a major new competitor, or a sharp rise in interest rates, may trigger an assessment of a model's current predictive accuracy) needs to be undertaken. In addition, a maintenance schedule to periodically test whether a model is still accurate, along with criteria to determine the point at which a model needs to be "refreshed" (i.e., rebuilt using more recent data), needs to be developed.

At the end of the project, the project team needs to *produce a final written report*. Depending on the nature of the project and its deployment, the report may be a summary of the project and the lessons learned from undertaking the project, or it may be a comprehensive presentation of all aspects of the project, with a detailed explanation of the data mining results. In addition to the written report, there may also be a final presentation.

The *project review task* is an assessment of what went both right and wrong with the project. Its main purpose is to provide the basis for learning about what should be done in a similar fashion, and what should be done differently, in future projects.

2.2.4 The Typical Allocation of Effort across Project Phases

A natural question to ask is the relative amount of time that is likely to be devoted to different phases of a project. Based on our experience, and those of others, we think reasonable guidelines are:

1. Business understanding: 5 to 15 percent

2. Data understanding: 5 to 10 percent

3. Data preparation: 50 to 60 percent

4. Modeling: 5 to 15 percent

5. Evaluation: 5 to 10 percent

6. Deployment: 10 to 15 percent

Probably the most surprising things about this list is the large amount of time devoted to data preparation and the fairly small amount devoted to modeling.

The length of time spent on the data preparation phase should never be under-estimated. While preparing the data seems like it should be straightforward, there is always a real horror show lurking in one of the database tables you will be using, just waiting to be found when you least expect it. The botched Y2K conversion discussed earlier is a classic example of a data preparation horror show. Unfortunately, horror shows never seem to repeat themselves across projects, so each project has it own unique one, and you have no idea what it will be until you stumble upon it. The actual modeling takes remarkably little time because it typically doesn't involve much more then selecting the values of several model parameters, selecting a set of variables, and clicking on an OK button to estimate the model. When things go badly in the modeling phase, it is typically a data problem that is the real culprit, requiring a return to the data preparation phase to correct it.

Having provided some basic background in this chapter and the previous one, we are now ready to begin to get our hands dirty by working more directly with data.

Part II

Predictive Modeling Tools

PART II

Predictive Modeling Tools

Chapter 3

Basic Tools for Understanding Data

The primary objective of this chapter is twofold. The first objective is to present a number of tools that are useful for the data understanding phase of the CRISP-DM process model (Chapman et al., 2000). The set of tools we present in this chapter is not the complete set we present in the book. The chapters on linear and logistic regression will present additional visualization tools useful in this phase of a data mining project. We have elected to hold off on the presentation of these visualization tools since they will have greater value in the context of those chapters. The second objective of this chapter is to introduce you to R and the modified version of the R Commander that we will use as the data mining workbench in this book.

Before we can successfully apply tools to better understand our data, we first need to know more about the nature of variable data types. It turns out that how we apply tools to understand a variable depends on what type of a variable it is. "Measurement scales" is the term used to describe the properties of variables that define their type. Consequently, this chapter begins with an introduction to measurement scales and variable types. Following this are four tutorials on basic tools for understanding data. The first tutorial shows you how to load data contained in an Excel file (the file format used to hold a remarkably high percentage of many organizations' data, often inappropriately) as well as data in an R "package" into R. The second tutorial covers obtaining simple descriptive statistics about a data set as a whole, and about individual variables within that data set. The third tutorial covers tools to examine the distribution across records of a single variable (known as a frequency distribution, which is visually displayed using a histogram), while the fourth tutorial looks at a simple multivariate analysis tool known as a contingency table used to look at the relationship(s) between two or more variables. For the last three tutorials, the tutorial will both describe tools and show you how to apply these tools to an example data set using R Commander.

3.1   Measurement Scales

Customers, products, companies, and any other "object" are described by their attributes. Attributes may vary from one object to another or from one time to another. To measure attributes, we assign numbers or symbols to them. The attribute of "eye color" of the object "customer" can be assigned the value blue, brown, or green. The attribute of "age" of the object "customer" can be assigned values of 23 or 76 years, and so on. Age will vary over customers and over time, while eye color (or at least natural eye color) only varies over customers.

A useful and simple way to specify the type of attribute is to identify the properties of numbers that correspond to the underlying properties of the attribute. An attribute like age, for example, has many of the properties of numbers. It makes sense to compare and order customers by age, as well as to talk about differences in age or the ratio of ages of two different customers (e.g., a customer who is 25 is half the age of a customer who is 50).

The following properties of pairs of numbers or values are typically used to describe attributes:

1. Distinctness: $=$ or $\neq$

2. Order: $\leq$, $\geq$

3. Addition: $+$ and $-$

4. Multiplication: $\times$ and $\div$

With these properties, we can define four types of attributes: *nominal, ordinal, interval,* and *ratio.* Each type possesses all of the properties of the preceding type. That is, if you can divide two numbers, you can also determine if they are equal or not. The reverse is not necessarily true—blue is not equal to brown, but blue divided by brown makes no sense. Thus, *nominal* variables only have the property of distinctness; *ordinal* variables also have the property of distinctness, but they possess the property of order as well; *interval* variables have the first two properties and the property of addition, so that we can meaningfully measure the difference between different values of an interval variable; and *ratio* variables possess all four properties, so we are able to say that one value of a ratio variable is three times as large as another value. Each variable type also has certain statistical operations that are valid for it. For example, it makes sense to talk about the average age of our customers, but not about their average eye color. Nominal and ordinal variables are also typically referred to as *categorical,* while interval and ratio variables are referred to as

Table 3.1: A Summary of Attribute Measurement Scale Types

Attribute Type		Description	Examples	Operations
Categorical	Nominal	Values are just different names. They provide only enough information to distinguish one object from the other (= or ≠).	Zip/postal codes, employee ID, eye color, gender	Mode, contingency table, chi-squared
	Ordinal	Values provide enough information to order objects as first, second, third, and so on $(<,\leq,>,\geq)$.	Hardness of minerals, grades, street numbers, gold–silver–bronze	Median, percentile, rank correlation
Numeric	Interval	The differences between values are meaningful $(+, -)$.	Calendar data, temperature in Celsius or Fahrenheit	Mean, standard deviation, Pearson's correlation, T-test
	Ratio	Both differences and ratios are meaningful.	Money, age, height; temperature in Kelvin	Geometric mean, percent variation or elasticity

numeric or *quantitative*. Table 3.1 summarizes attribute measurement scale types.

We also use the terms *discrete* and *continuous* to describe attributes. Continuous attributes can take on any real number values—such as 36.35 years old. Discrete attributes do not take on these sorts of intermediate values. Brown and blue are discrete, as are first, second, and third. Buy or not buy, which only has two levels, is known as a *binary* discrete variable (or simply a *binary variable*). Usually, interval and ratio variables are continuous, and ordinal and nominal variables are discrete, hence *continuous* and *discrete* are often used interchangeably with *numeric* and *categorical*. However, there are exceptions to this. For example, count data (such as the number of separate purchase occasions a customer has in a particular store in a given year) is both discrete and ratio scaled since the number of trips for any particular customer will always be an integer value.

3.2   Software Tools

The software tools used with this book are written in R (R Development Core Team, 2011), a language and system for computational statistics. R is open source software, developed by a large, worldwide team of developers. The software consists of a base system and an extensive library of add-on packages. A recent survey (Rexer et al., 2010) found that R was the software product that had the largest percentage of users (43%) of data mining professionals.

On both Windows and OS X, R has a limited graphic user interface (or GUI). However, these GUIs are operating system specific, and both are fairly difficult to extend. Instead, there are R bindings to several GUI toolkits, including Tcl/Tk, GTK2, Java Swing, and GTK2. Using these GUI tookits, Deducer (Fellows, 2011), pmg (Verzani, 2011), and the R Commander (Rcmdr; Fox, 2005) offer fairly complete basic GUIs for R, and all three are extensible. In addition, the rattle package (Williams, 2009) offers a GUI to R that is oriented toward data mining. One thing that separates the various R GUI packages is the GUI toolkit they are based on. In turn, these different toolkits have different installation requirements, with some requiring more effort to install than others. The GUI toolkits also differ in the level of "quirks" they exhibit across operating systems.

The easiest GUI toolkit to deal with from a user installation perspective, and the one that is most mature, is Tcl/Tk. The Tcl/Tk toolkit is installed as part of an R installation on Windows (so if you have R, you have the Tcl/Tk toolkit on a Windows system), it requires only one additional software package (beyond R itself) to be installed on OS X, and is likely to already be installed on most Linux systems. The only fairly complete basic R GUI that uses the Tcl/Tk toolkit is the R Commander. The combination of this fact, along with the R Commander's greater level of maturity, and the ease with which custom

Figure 3.1: The R Project's Comprehensive R Archive Network (CRAN)

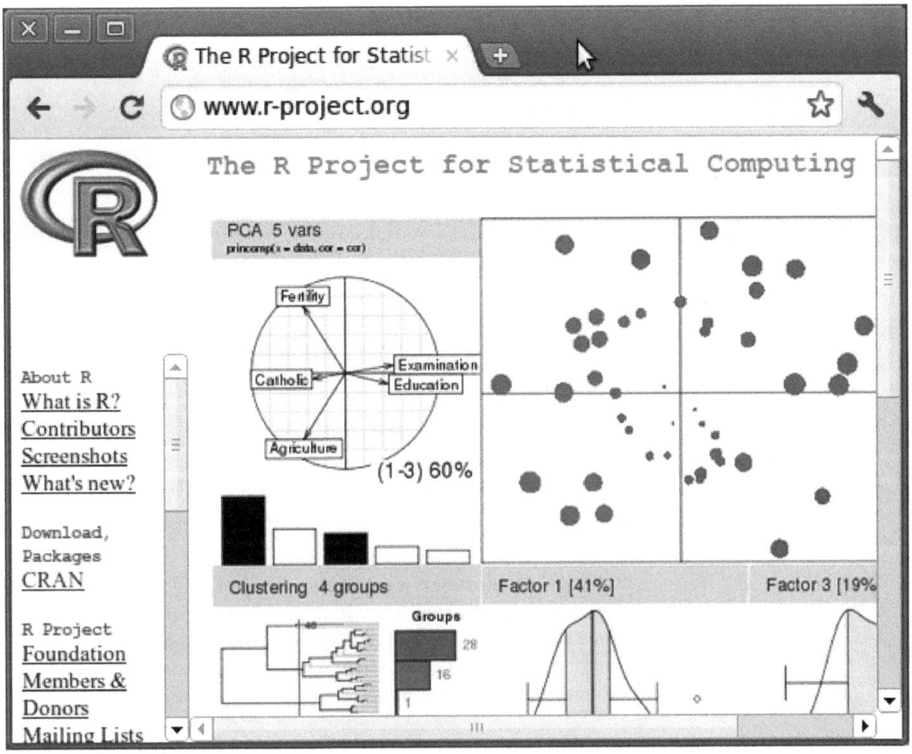

plug-in packages can be developed led us to adopt it for the software tools to accompany this book.[1]

The R system software as well as most R packages are available from the Comprehensive R Archive Network (which is typically referred to as CRAN). CRAN has mirror repositories located around the world. Selecting the mirror repository nearest to you should reduce the time it takes to download the software. Like R itself, the software developed for this book is open source, and is publicly available from CRAN. In what follows, we provide step-by-step instructions on how to obtain and install the needed software on both Windows and OS X systems.

3.2.1 Getting R

The preferred way to obtain R is to open a web browser and go to the R project's main site at http://www.r-project.org, where you will see a page that looks similar to Figure 3.1. In the left-hand navigation bar you should

---

[1]The data mining–oriented rattle package uses the GTK2 toolkit, which makes it potentially challenging to install on Windows and (particularly) OS X systems. In addition, our approach to applied data mining is a bit different from the one implicit in the rattle GUI.

Figure 3.2: The Entry Page into the Comprehensive R Archive Network (CRAN)

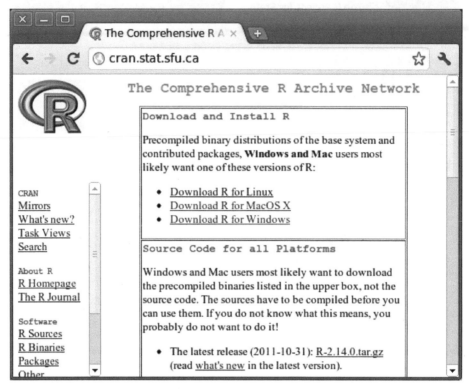

see a link to CRAN (which has been placed in a rectangle in Figure 3.1). Click on this link, which will bring you to a page to select the CRAN mirror site (organized by country) from which to download the software. After selecting a mirror site you will see the page shown in Figure 3.2. At the top of this page are links to operating specific precompiled binaries. Select the link to the operating system running on your computer.

After clicking on the appropriate link, Windows users will see the page shown in Figure 3.3. Click on the "base" subdirectory (the top-left link on the main part of the page) to get to the actual download page, which is shown in Figure 3.4. Click on the top link of the main part of this page to download the installer for the current version of R. You may also want to browse through the frequently asked questions to see how to troubleshoot issues that may come up during the installation (such as administrator privilege issues under Windows 7 and Windows Vista).

Mac OS X users will first see the page shown in Figure 3.5. Click on the link in the top-left of the main page to download the installer for the base R system. In addition, you will need to click on the link to "the tools directory" to get

Figure 3.3: The R for Windows Page

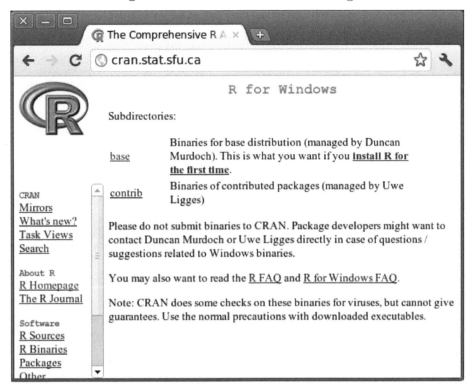

Figure 3.4: R for Windows Download Page

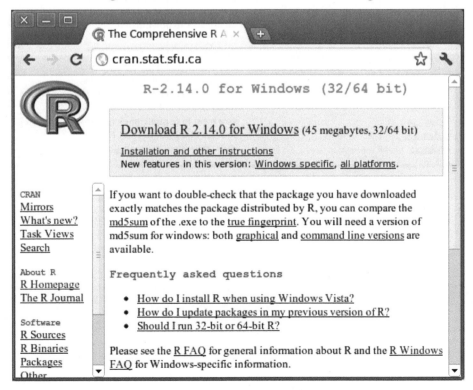

Figure 3.5: The R for Mac OS X Download Page

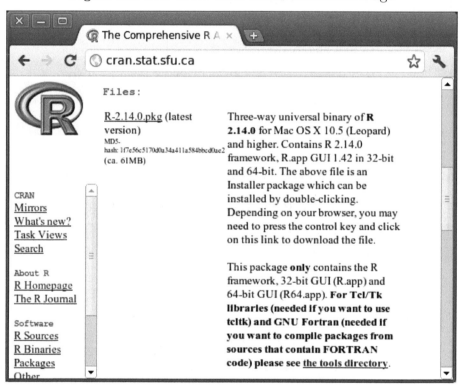

Figure 3.6: Mac OS X X11 Tcl/Tk Download Page

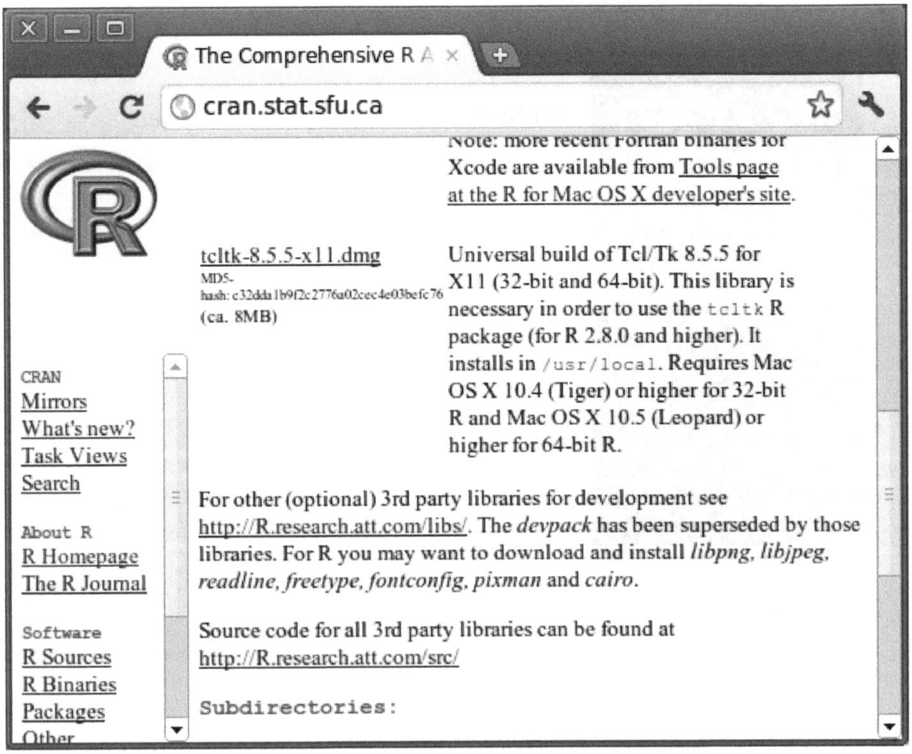

the needed Tcl/Tk library. On the tools directory page (shown in Figure 3.6), scroll down to the Tcl/Tk for X11 section, and click on the link to download the *.dmg file of this library.

3.2.2 Installing R on Windows

Navigate to where you save the R for Windows installers. Double-click on the installer to bring up the installation wizard, the first page of which is shown in Figure 3.7. Click on the "Next" button in this window, and accept the default setting until you come to the "Startup options" window (shown in Figure 3.8). In this window select the "Yes (customized startup) option, then press "Next," and the "Display Mode" installer window shown in Figure 3.9 will appear. In this window select the "SDI (separate windows)" options. At this point, go through the remaining windows of the install wizard, accepting the default options. After you have completed the wizard, R will be installed on your system. By default, a desktop icon to launch R will be added.

Figure 3.7: The R for Windows Installation Wizard

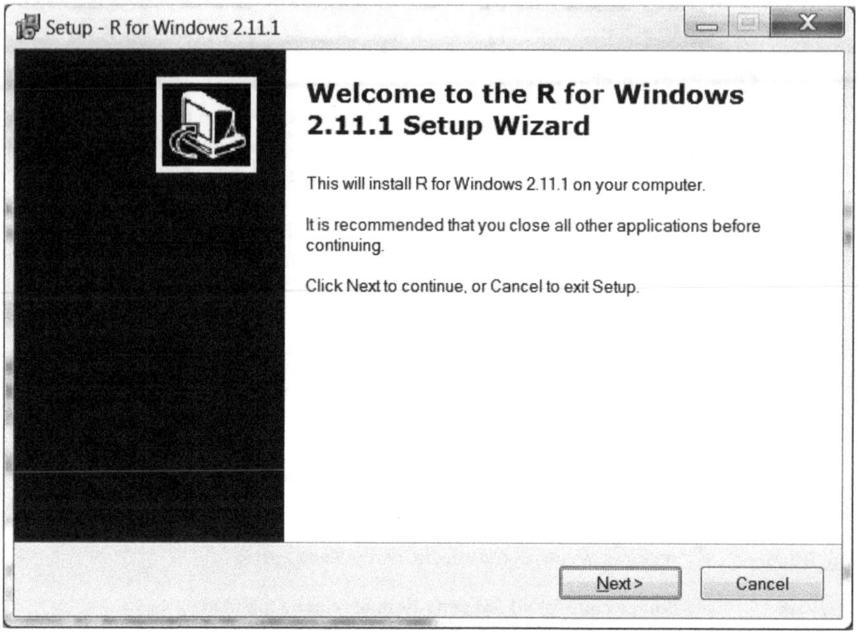

Figure 3.8: The Customized Startup Install Wizard Window

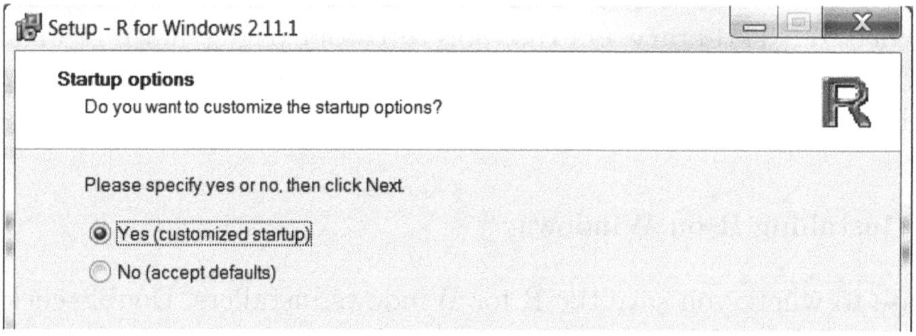

Figure 3.9: The Installation Wizard Display Interface Selection Window

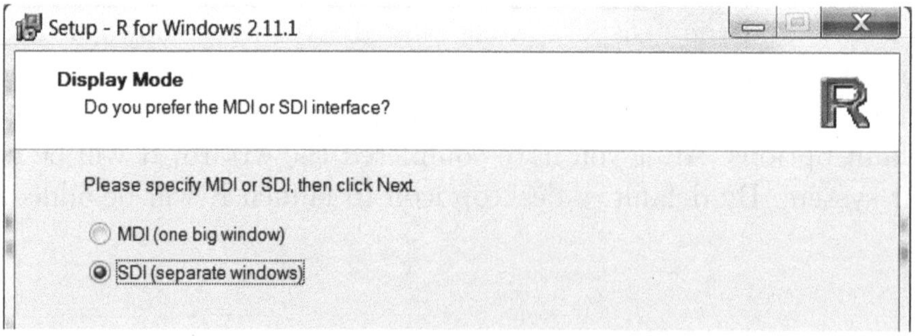

Figure 3.10: The R for Mac OS X Installer Wizard Splash Screen

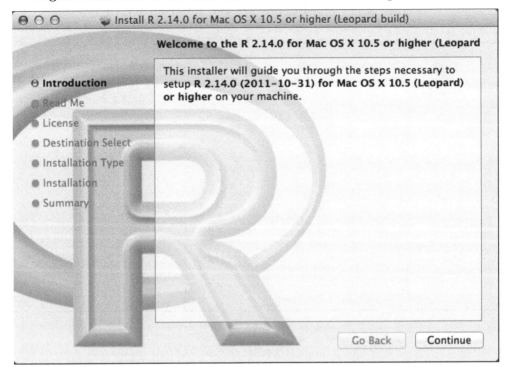

3.2.3   Installing R on OS X

For reasons of convenience, you may want to copy the two files you downloaded onto the desktop. To start the installation, double-click on the R-2.x.x.pkg (the x's will actually be replaced by numbers), which will launch the R installer wizard, the first window of which is shown in Figure 3.10. In the case of OS X, accepting all of the default options works well. As a result, just follow the instructions in the installation wizard. Along the way you will be asked to accept the license terms (which you should) and asked to authenticate (i.e., enter your password) just before the software is actually installed on your system. The next thing to do is double-click on the tcltk-8.5.x.dmg file (again, the x will be replaced by doing this). The tcltk-8.5.x.dmg is a disk image file, so double-clicking on it will cause the disk image to uncompress. This will reveal a single file (see Figure 3.11), which will have the name tcltk.pkg. Double-click on this file to start the installer, and you should see a window similar to the one in Figure 3.12. You should accept all the default settings for this package as you go through the wizard, and you will again be asked to accept the license and enter your password to authenticate just prior to the actual installation of the software.

Figure 3.11: The Uncompressed Tcl/Tk Disk Image

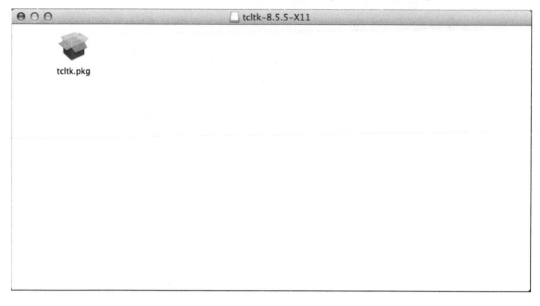

Figure 3.12: The Tcl/Tk Installation Wizard

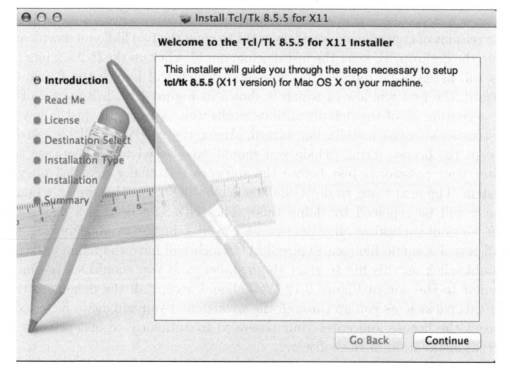

Figure 3.13: The Source Command to Install the RcmdrPlugin.BCA Package

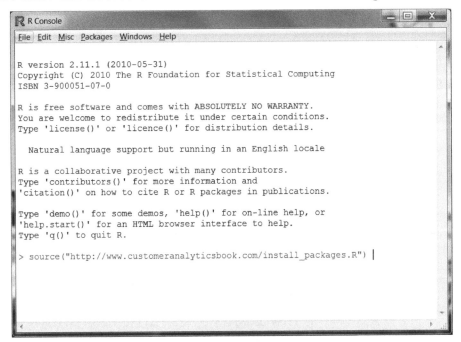

3.2.4 Installing the RcmdrPlugin.BCA Package and Its Dependencies

To install the software that goes along with this book you need to launch R. Mac OS X users will want to launch X11 prior to launching R. X11 should live in the Utilities folder within the Applications folder of the Finder. Windows Vista and Windows 7 users are likely to want to launch R by right-clicking on the R desktop icon and selecting "Run as Administrator." Once R is launched, you should see the R Console window (the Windows version of the R Console is shown in Figure 3.13). The R console provides a command line interface to R, along with a small number of GUI tools to work with the R system. To install the RcmdrPlugin.BCA package, along with the packages that it uses, we need to issue one command using R's command line interface. The R prompt is a greater than sign (>), which will be at the bottom of the R Console. At the prompt enter the command: source("http://www.customeranalyticsbook.com/install_packages.R"), which is illustrated in Figure 3.13, and then press enter. At this point a pop-up window will appear (shown in Figure 3.14) asking you to select a CRAN mirror repository from which to download the needed packages. Select the mirror site nearest to you, and then press OK. At this point you will see a lot of activity in the R Console window as R downloads and installs the needed packages. When R is done installing the software, a new prompt (a ">") will appear at the bottom of the R Console. When it does, you can launch into

Figure 3.14: Selecting a CRAN Location for Package Installation

Figure 3.15: The R Commander Main Window

the data mining tools that accompany this book by entering the command: library(RcmdrPlugin.BCA). What will happen next is the R Commander GUI will appear, momentarily disappear, and then reappear. When it reappears, you should see a window very similar to the one shown in Figure 3.15. In future sessions, you will need to enter the "library(RcmdrPlugin.BCA)" command at the beginning of the session to load the needed tools.[2]

---

[2]For Mac OS X users, if you enter this command and receive an error message that ends with the line "Error: package 'tcltk' could not be loaded," it means that you did not start X11 before launching R. Exit from R using the File → Quit drop-down menu, start X11, start R, and then enter the library command to load the RcmdrPlugin.BCA package.

3.3   Reading Data into R Tutorial

Most of the data sets we will be using in the book are included with R Commander. Others will be imported from outside sources. R Commander is able to directly import a variety of data formats. Even if you have data in a format that R Commander does not recognize, you should be able to convert it using the application that generated it to export it in a format that R does recognize, and this will give the flexibility to easily use almost any data format you happen to have. In this lab you will learn (1) how to convert an outside file and import it, and (2) how to access data from an R data library so that you can work with it through R Commander. For the first task we will convert an Excel file to a "comma separated values" or CSV file and import that file. (Under Windows, R Commander can read an Excel file so this would not normally be done, but for OSX, Linux, and other operating systems, the ability to read Excel files is not readily available). We will begin by converting and importing the Jack and Jill Clothing Company children's apparel spending data set from an Excel workbook.

1.

Use a web browser and go to http://www.customeranalyticsbook.com/jackjill.xls, and you will then be asked whether you want to open or save this file. Choose Save to download the file to your local drive.

2.

The **Save As** dialog box will appear asking you where to save the file "jackjill.xls" (the suffix may not appear, depending on your operating system's settings). Navigate to the folder in which you would like to keep your files related to the tutorials for this book, and then press the **Save** button.

3.

After the file "jackjill.xls" has been downloaded to your local drive, locate it and **open it in Excel.** Once you have done this, you should see the file shown in Figure 3.16. This file contains children's apparel spending and household socioeconomic information for 557 households for the year 1992. Look through the file. You will notice that most of the columns contain text as opposed to numbers. Most of the variables in this data set are categorical (some nominal and some ordinal), and the text values describe the category.

Figure 3.16: jackjill.xls

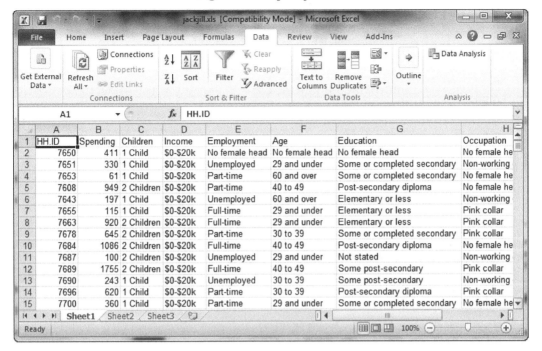

R Commander can read several different types of files, including Excel files like this one. However, if a file cannot be read, often changing to a different file type will solve the problem. To illustrate this first trick in your tool kit, we will convert the *.xls file to a comma separated value file, with the file extension *.csv.

4.

As shown in Figure 3.17, use the pull-down menu command **File→ Save As**, which will bring up the **Save As** dialog box shown in Figure 3.18. Navigate to where you want to save the CSV file you are about to create, and, as shown in Figure 3.18, use the drop-down menu to **select "CSV (Comma delimited)(*.csv)" and save** the file. By default, the file will have the name jackjill.csv.

5.

Launch R by double-clicking its icon, which, after a few moments, will bring up the R console. **Enter the command library(RcmdrPlugin.BCA)** to bring up the R Commander GUI, which is shown in Figure 3.19. It contains drop-down menus, buttons, a script window, an output window, and a messages window.

Figure 3.17: Saving a File to Another Format in Excel

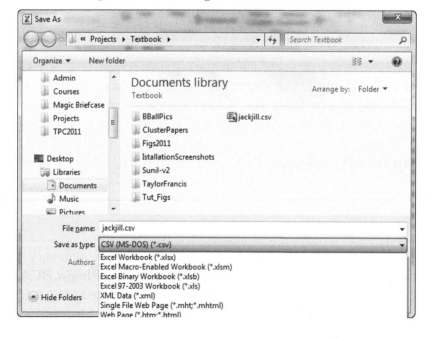

Figure 3.18: Saving a CSV File in Excel

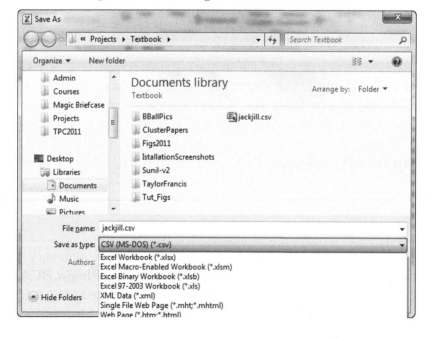

Figure 3.19: Importing Data into R

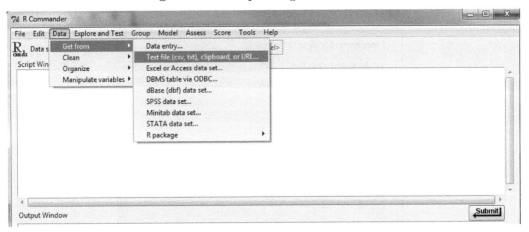

6.

In R Commander use the pull-down menu command **Data** → **Get From** → **Text file (csv, txt) clipboard, or URL...**, which will cause the dialog box in Figure 3.20 to appear.

Figure 3.20: The Import Text File Dialog Box

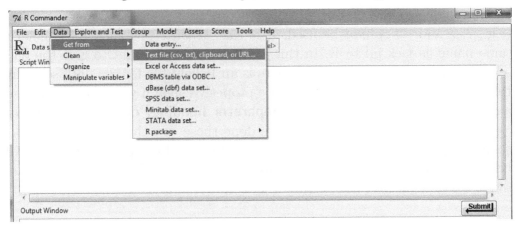

7.

In this dialog box, **enter jack.jill** in the "Enter name for data set:". Whatever name you enter here is what R will use to refer to this data set after it is read into R and stored as an R readable file. It does not have to be the same as the name of the input file, which will be the case here. Generally, you have a wide degree of latitude in what you can name a data set. Although

Figure 3.21: The Completed Import Text Dialog Box

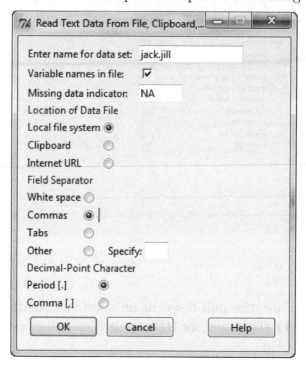

the name needs to start with a letter, it *can* contain letters, numbers, periods (.), and underscores (_), but other characters that are not numbers or letters (e.g., ?, !, and <) *cannot* be used. *R is case sensitive*, so Jack.Jill is not the same name as jack.jill to R. In this instance, you will want to give the data set the exact name jack.jill since there is an R help file for the data set using this name, and the help file will not be properly called if you give the data set another name. Next, in the **Field Separator heading, click on "Commas"** as the field seperator. Once you have done these two things, the dialog box should appear as it does in Figure 3.21. Once it does, press **OK**. This will bring up a standard **Open File** dialog box shown in Figure 3.22. Find and select the new file you have created, jackjill.csv.

8.

The bottom window in R Commander provides informative messages, such as errors, and should always be monitored. If the file has imported correctly, this window will now show that the dataset jack.jill has 557 rows and 9 columns, indicating successful loading of the database. If you press the **View data set** button on the R Commander toolbar you will be able to view the jack.jill data set, as is shown in Figure 3.23.

Figure 3.22: The Standard Open File Dialog Box

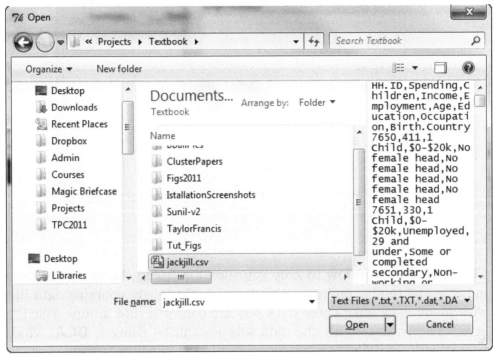

Figure 3.23: Viewing the jack.jill Data Set

	HH.ID	Spending	Children	Income	Employment	Age	
1	7650	411	1 Child	$0-$20k	No female head	No female head	
2	7651	330	1 Child	$0-$20k	Unemployed	29 and under	Some or comp
3	7653	61	1 Child	$0-$20k	Part-time	60 and over	Some or comp
4	7608	949	2 Children	$0-$20k	Part-time	40 to 49	Post-se
5	7643	197	1 Child	$0-$20k	Unemployed	60 and over	Ele
6	7655	115	1 Child	$0-$20k	Full-time	29 and under	Ele
7	7663	920	2 Children	$0-$20k	Full-time	29 and under	Ele
8	7678	645	2 Children	$0-$20k	Part-time	30 to 39	Some or comp
9	7684	1086	2 Children	$0-$20k	Full-time	40 to 49	Post-se
10	7687	100	2 Children	$0-$20k	Unemployed	29 and under	
11	7689	1755	2 Children	$0-$20k	Full-time	40 to 49	Some
12	7690	243	1 Child	$0-$20k	Unemployed	30 to 39	Some
13	7696	620	1 Child	$0-$20k	Part-time	30 to 39	Post-se
14	7700	360	1 Child	$0-$20k	Part-time	29 and under	Some or comp

Figure 3.24: Reading a Data Set in a Package

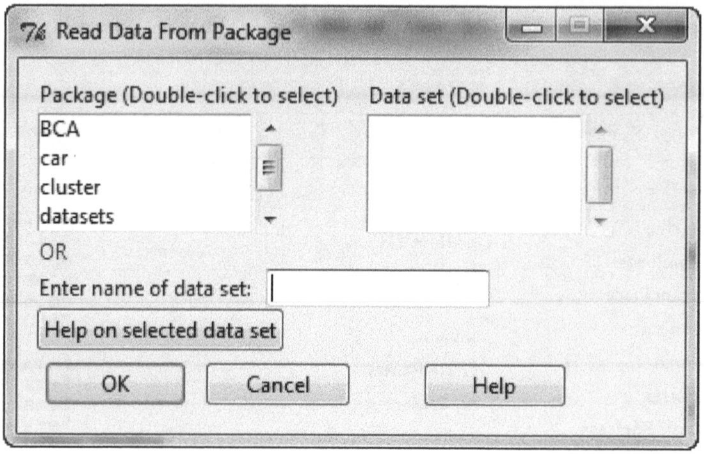

9.

Close the view window to keep your desktop from getting too cluttered. As indicated at the beginning of this tutorial, R already contains data libraries with many data sets. These data sets are collected into groups called "packages." The package with the data sets we will be using is **BCA**, which was loaded with the RCmdr software when the RCmdr library command was initially entered.

10.

Select **Data → Get From → R → package Read data set from an attached package...**, which will bring up the dialog box shown in Figure 3.24, and shows the packages currently loaded.

Double-click on the package BCA to list the five data sets available in the package in the right box. Scroll down to CCS, and double-click on it (Figure 3.25). Press **OK**. **View this data set** as in step 9. Close the data set view window. Note that the bottom window shows that the data set CCS has 1600 rows and 20 columns, and that the Data Set button below the main menu has changed to **CCS**, indicating that this is now the "active" data set.

11.

Re-activate jack.jill by **clicking on the Data Set button**—which now has CCS on it—and then **select jack.jill** from the dialog box. After closing the box, jack.jill will be the active data set.

Figure 3.25: Selecting the CCS Data Set2

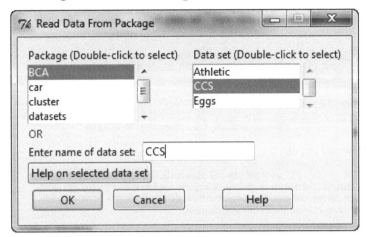

Help Files

Examine the help documentation for the jack.jill data set by selecting **Data → Clean → Help on active data set (if available)** from the R Commander pull-down menus. This will launch a web browser with the help file displayed. If you did not give the data set the exact name "jack.jill," you will get a warning message indicating that no documentation for the name you gave the data set exists. If this is the case, repeat steps 7 and 8 taking care to give the data set the name that corresponds to the help file. After reading the help file for the jack.jill data set, close the help system window.

12.

You now have two data sets read in and available to work with, jack.jill and CCS. To save them so that they will be available to you for the next tutorial, select **File → Save R workspace as. . .**, which will bring up a "Save as" dialog box. Navigate to the folder where you wish to save the workspace file, such as the Desktop. Workspaces have the filename extension ".RData" so **type Tutorial3.3.RData** as shown in Figure 3.27. Select **Save**. The workspace file will save your data and (later on in the book) any models you have constructed. It will also allow you to start working again where you have left off; to send files to another computer running Windows, Mac OS X, or Linux; and to share analysis work with the members of your team. When you wish to use a saved workspace file, make sure R is not running, and then start up R and R Commander by double-clicking on the workspace file icon.

Figure 3.26: Data Set Help

Figure 3.27: Saving a *.RData File

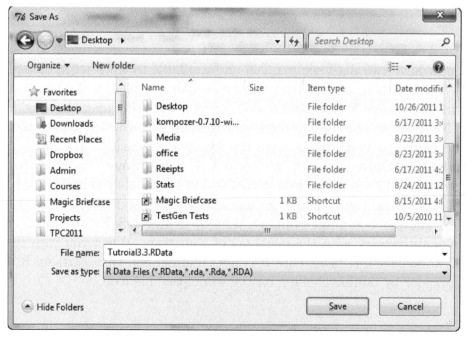

13.

At this point you can quit from the R Commander and R by selecting **File** →
Exit → **From Commander and R**. R Commander will present a dialog box
asking if you really want to exit. If you select **OK**, a series of dialog boxes will
appear asking for confirmation and if you want to save the script and output
files of the session. The script file contains a record of all the commands you
issued during this R session. Saving it allows you to instantly duplicate the
entire session later. We will not need the script file, so select **No**. The output
file is not very interesting for this tutorial so also select **No** for it.

3.4   Creating Simple Summary Statistics Tutorial

In this tutorial you will see how to use several different tools that will allow you
to quickly get an initial understanding of the data you are examining. If you
are continuing from the last tutorial, you should be ready to go. Otherwise,
double-click on the Tutorial3.3.Rdata file you created as part of the last
tutorial. (Do not start R from its own icon when you want to start from a saved
workspace.) **Enter library(RcmdrPlugin.BCA)** in the R Console to Start R
Commander. Make jack.jill the active data set by using the **Data Set** button.

1.

There is one "data cleaning" chore that should be done for most data sets.
The jack.jill data set contains the variable HH.ID, and is an example of a case
identifier (or observation identifier). For the jack.jill data set, it is a household
identification number. Since these "variables" are merely identifiers, they will
not actually be used in any analyses that we will do. We can ensure that
we will never accidentally use it as an analysis variable by setting HH.ID as
a record-name data type now. This step should ALWAYS be taken with a
new data set that has identification variables — and most will. Click on **Data**
→ **Clean** → **Set record names...**, which bring up the dialog box shown in
Figure 3.28.

2.

Within this dialog box **scroll to and click on HH.ID** with your mouse to
highlight it, and then press **OK**. HH.ID has now become an observation identifier, and no longer appears as a variable in jack.jill. You can check this by

Figure 3.28: The Set Record Names Dialog

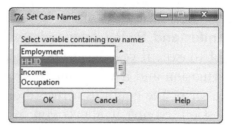

Figure 3.29: Variable Summary for the jack.jill Data Set

```
> variable.summary(jack.jill)
                 Class %.NA Levels Min.Level.Size    Mean      SD
Spending       integer    0     NA              NA 784.377 761.6432
Children        factor    0      4              15     NA      NA
Income          factor    0      8              42     NA      NA
Employment      factor    0      4               9     NA      NA
Age             factor    0      6               5     NA      NA
Education       factor    0      7               2     NA      NA
Occupation      factor    0      6               8     NA      NA
Birth.Country   factor    0      6               7     NA      NA
```

clicking the **View data set** button. Close the data set window after viewing to reduce desktop clutter.

3.

To get information on all of the variables in this data set, select **Data → Clean → Summarize Variables**, which will cause the table in Figure 3.29 to appear in your R Commander Output window.

Variable Attributes

This table lists a number of attributes for each variable in the data set. Inspecting this table is a critical part of the data understanding phase, and helps with identifying data problems and hence the data preparation steps necessary to fix them.

The variable names appear in the first column. The second column (Class) gives the class of each variable, which incorporates both the measurement scale and the internal computer coding used for the variables. The combination of scales and coding can initially be confusing, but there are only four classes, and getting comfortable with them is essential for data analysis. The possible classes of variables are *numeric, integer, factor,* and *character. Numeric* and *integer* variables are ratio-scaled numbers (integer variables do not have fractional values, but for our purposes can be treated as numeric variables). A *factor* is a categorical variable. The summary table does not distinguish between ordered (ordinal-scaled) and unordered (nominal-scaled) factors, but we will usually treat them as nominal. From this, you can conclude that we will be working primarily with ratio and nominal variables. A *character* variable only contains text as labels and are not variables that can be used in an analysis. This is fine if the variables are merely record identifiers, but if they are variables that need to be used in an analysis, we will need to convert them. We will see how to convert them later, usually to factors.

The third column (%.NA) gives the percentage of values that are missing for that variable. In this instance, none of the variables has missing values. However, in most database marketing applications you will encounter data sets that have variables with many missing values. An essential data cleaning step in the data preparation phase of a project will be to decide what, if anything, needs to be done about missing values. We will leave this important question for later.

The third and fourth columns (Levels and Min.Level.Size) provide information for factor variables (thus `SPENDING,` a numeric variable, has no information, `NA`, in this column). The Levels column indicates the number of categories or "levels" for a factor variable. For example, a factor "gender" would have two levels, male and female. Min.Level.Size indicates the number of records for the category or "level" that has the fewest records. A data set consisting of 14 records, with 11 males and 3 females, would have the a minimum level size of 3 for gender. These columns are provided because factors with many levels and levels with too few records can cause problems during the model building phase. Moreover, any observed effects associated with levels with few records are statistically unreliable. What qualifies as too few levels depends on the analysis details, but as a start we should at least note that `EMPLOYMENT, AGE,` `EDUCATION, OCCUPATION,` and `BIRTHCNTRY` are all likely cases. One common thing to do in the data preparation phase is to combine two or more factor levels into one level, which thus has more records and increases the minimum level size, and which also decreases the number of factor levels. In the next tutorial you will see exactly how to do this using R Commander.

Exercise: How many categories for age are defined in these data? The age category with the fewest households has how many households?

The fifth and sixth columns (Mean and SD) contain the mean and standard deviation for each numeric variable in the data set (the values are missing for factor and character variables). The mean is useful as an indication of the magnitude of a variable (in the tens, hundreds, thousands, etc.). The standard deviation provides a measure of the spread or range of the data. One thing to be on the lookout for is a variable with a standard deviation equal to zero. A variable that has a standard deviation of zero is not to be a variable at all, it is a constant (i.e., it only has one single value in all records), and the inclusion of this variable as input to a number of data mining methods can

Figure 3.30: The Numerical Summary Dialog

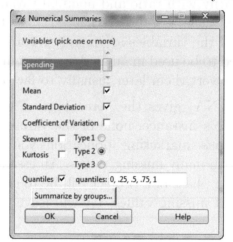

result in problems. Generally, we either delete constants from the data set, or determine why the variable is a constant since it suggests that there may have been a problem in prior processing of the data.

4.

In the next tutorial we will see a number of tools to get descriptive statistics for factor variables. For the remainder of this tutorial we will look at simple summary statistics for numeric variables. We can gain some additional information about the SPENDING variable by selecting **Explore and Test →Summarize → Numerical summaries...**, which will generate the dialog box shown in Figure 3.30.

5.

The only numeric variable in jack.jill is SPENDING, so it is the only variable shown in the **Variables (pick one or more)** selection box. To make sure it is properly selected, click on the variable name. We will leave the other options as they are. Right now we will only do a summary of SPENDING across all households, so press **OK** in order to create the output table shown in Figure 3.31.

You will notice that the mean and standard deviation ("sd" in the new output) is the same as it is in Figure 3.29 (761.6432). The additional information is the values of the variables at each of its quartiles. The 0% quartile is the minimum value of SPENDING in the data set ($13), the 50% quartile is the median value of SPENDING ($585), and the 100% quartile is the maximum value of SPENDING ($5940) in the data set. The large gap between the 75%

Figure 3.31: A Numerical Summary of SPENDING

```
> numSummary(jack.jill[,"Spending"], statistics=c("mean", "sd", "quantiles"),
+   quantiles=c(0,.25,.5,.75,1))
    mean       sd 0% 25% 50% 75% 100%   n
 784.377 761.6432 13 309 585 965 5940 557
```

and 100% quartiles (relative to the other inter-quartile gaps) indicates that
the distribution of SPENDING is likely to be highly skewed, with a very small
percentage of households having extremely high spending levels. The fact that
the mean is much higher than the median is also a result of such skewing. The
last value ("n") in the summary is the number of non-missing values for the
variable. Since the data set contains 557 records, this again confirms that
SPENDING has no missing values.

Exercise: Return to the numerical summaries dialog box, but this time
select the Summarize by groups... button. This will bring up a dialog
box that lists the categorical variables. In this box, **select CHILDREN,** then
OK, and then OK again in the summary box. Compare the output with the
previous output, and explain what you see.

6.

Because there is only one numeric variable in the jack.jill data set, it is not
useful for the next set of tools we want to illustrate. Consequently, at this
point load the CCS data set. To do this, press on the **Data Set** button and
select **CCS** as the active data set. In a later tutorial we will describe the CCS
data set in much greater detail. For now it is important to know that this
data set relates to fundraising activities undertaken by a Canadian Charitable
Society (or CCS). Some of the variables in the data set capture past behavior
on the part of individuals with respect to their giving to the CCS, while
other variables are derived from Census data and are averages of demographic
attributes in the individual's neighborhood.

7.

At this point we want to gain an understanding of how the variables that
capture past giving behavior relate to one another. In particular, we want to
look at the correlations between these variables. Select **Explore and Text** →
Summarize → **Correlation matrix...** to generate the dialog box shown in
Figure 3.32.

Figure 3.32: The Correlation Matrix Dialog

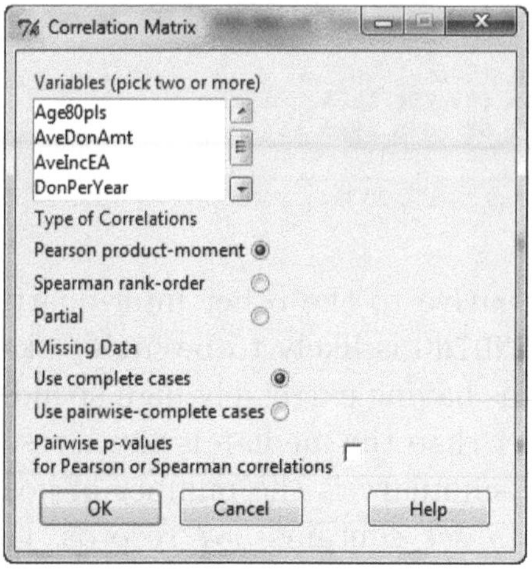

Figure 3.33: Correlation Matrix Results

```
> cor(CCS[,c("AveDonAmt","DonPerYear","LastDonAmt","YearsGive")],
+    use="complete")
            AveDonAmt DonPerYear LastDonAmt    YearsGive
AveDonAmt  1.00000000  0.1262545 0.86586622  0.03632848
DonPerYear 0.12625451  1.0000000 0.09886880 -0.45153640
LastDonAmt 0.86586622  0.0988688 1.00000000  0.05135673
YearsGive  0.03632848 -0.4515364 0.05135673  1.00000000
```

8.

The four numerical variables that capture past giving behavior are `AveDonAmt` (the average amount given per donation to the CCS by a giver), `DonPerYear` (the number of donations per year made by a giver), `LastDonAmt` (the amount of a giver's last donation), and `YearsGive` (the number of years since the giver's first donation to the CCS). Within the **Variables (pick two or more)** selection box, **select these four variables**. You can select multiple variables by pressing the control key (Ctrl on most keyboards) when clicking on a variable with a mouse. Once you have selected these four variables, press the **OK** button to produce the correlation matrix shown in Figure 3.33.

Correlations

The correlation matrix indicates that `AveDonAmt` and `LastDonAmt` have a correlation of 0.866, which is high (correlation coefficients are bounded between -1 and 1). It is not surprising that the higher the *average* donation amount, the higher the *last* donation amount. In subsequent tutorials on predictive modeling, we will see that using two highly correlated (either postively or negatively) variables such as these as predictors can cause problems, and that we need to be aware of and manage those problems. The other two variables that are somewhat (negatively) correlated are `DonPerYear` and `YearsGive`, with correlation −0.452: the more donations per year, the less time the individual has been a donor, perhaps somewhat unexpected. This correlation is probably not strong enough to cause problems for predictive models using both of these variables as predictors.

Exercise: Mr. White donates to the society three times each year. Mrs. Brown donates twice each year. Who is more likely to have donated first?

At this point you can either quit from R and R Commander, or go directly to the next tutorial. If you quit, you may save the R workspace as before. Note that in this tutorial we have not altered either the jack.jill or CCS data sets. Therefore, as long as you saved the R workspace after the "Reading Data into R Tutorial," there will be no change and there is no need to save the R workspace again.

3.5 Frequency Distributions and Histograms Tutorial

In this tutorial you will learn to use R to explore one variable at a time, called *univariate* (one variable) analysis. The two tools are frequency distributions and histograms. We will also learn a handy tool for converting between variable types. Frequency distributions and histograms are intended to be used with categorical variables (i.e., variables that have either a nominal or ordinal scale), which R calls "factor variables." The example used in this tutorial is the amount of money spent on children's apparel by 557 households for a one-year period, which is a ratio-scaled continuous variable that we will learn how to convert into a categorical variable. This will be useful in the future as there is often a need to break up the range of a continuous variable into a number of discrete categories. This is often referred to as "binning" or "bucketing" a continuous variable, and R Commander provides a very nice tool for this.

1.

If you did not quit R and Rcmdr after completing the last tutorial, continue with the next step. If you did quit, restart R by **double-clicking on the workspace file Tutorial3.3.RData**. If you saved it on your desktop, it will appear with a blue R icon.

Figure 3.34: Select Data Set Dialog

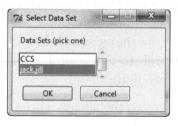

Figure 3.35: Histogram Dialog

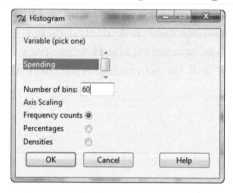

2.

Once the R Commander window is up, click on the data set button (the button will have either the label **No active data set** or **CCS** on the R Commander toolbar). The button brings up the **Select Data Set** dialog box shown in Figure 3.34. The two data sets read into R from the first tutorial are available. Select jack.jill by **clicking on it, and then on OK**.

3.

We are ready to make a histogram of spending on children's apparel across households. To access R Commander's **Histogram** dialog box, use the pull-down menu **Explore and Test** $\rightarrow$ **Visualize** $\rightarrow$ **Histogram...**, which brings up the **Histogram** dialog box shown in Figure 3.35.

4.

As you have probably guessed, Figure 3.35 is the "filled-in" dialog box, with the number of bins set to 60, while the one on your screen is set to **auto**. The only variable that appears in the variable list is SPENDING. The reason for this is that SPENDING is the only continuous (coded as "numeric') variable in the jack.jill data set. The histogram tool is only for continuous variables. For factor variables, a similar visualization tool is a bar graph. We will use the bar graph tool later in this tutorial.

5.

The variable **Spending** is automatically selected. The **Number of bins** field allows you to enter the number of bins (or "buckets") to use in breaking up the range of the SPENDING variable. The values for this variable range from $13 to nearly $6000, thus entering a value of 60 in this field will break the range into roughly $100 increments, and since we have 557 records, will give an average of about 9 records in each bin. This will give us a pretty good visual display of the distribution of children's apparel spending across the households in our sample. **Enter the number 60** in the "Number of bins:" field, and then press the **OK** button. R will create a graphics window on your desktop containing the histogram shown in Figure 3.36.

Exercise: By inspecting the histogram, can you predict whether the average or the median of spending will be higher? Explain, and describe each in terms of its application to ordinal and interval data.

Figure 3.36: Children's Apparel Spending Histogram

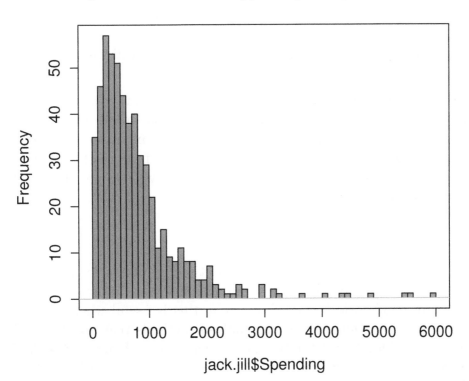

Figure 3.37: Save Plot Dialog

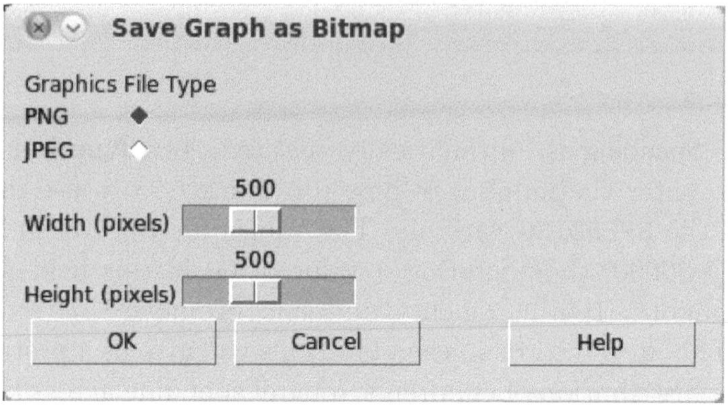

6.

Copying, Pasting, and Recording Graphics

Under Windows, a quick way to copy a graph to the clipboard for pasting into another document is to **right-click on the graph** and from the context menu select **Copy as a Metafile. . .** (or as a bitmap). Try this out now since it will be useful throughout the rest of the book. Open a Word document, and then paste the graph into the document. When you create you next plot, the current plot will be written over and lost. If you want to save all of your plots and be able to switch back and forth between them, **click on History → Recording**. In most of your analysis sessions, you will want to have plot recording turned on.

As an aside, under both Mac OS X and Linux, you can take advantage of the pull-down menu options under **Graphs → Save graph to file**. The best option is likely to be **Graphs → Save graph to file → as bit map. . .**, which will bring up the dialog box shown in Figure 3.37 (taken from the Linux version). Save the file in PNG format, press **OK**, and you will be presented with a standard **Save as. . .** dialog box.

7.

Having examined the histogram tool, it is time to take a look at the tool for "binning" a continuous variable. Binning converts a continuous (numeric) variable into a categorical (factor) variable; this is useful when you have a continuous variable, but wish to use analysis methods that are designed for categorical variables. Select the pull-down menu option **Data → Manipulate variables → Bin numeric variable. . .**, which brings up the **Bin a Numeric Variable** dialog box shown in Figure 3.38.

Figure 3.38: Bin Numeric Variable Dialog

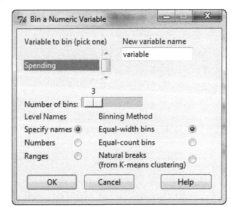

8.

Binning Methods

R Commander's binning tool offers three different methods of dividing a continuous variable into a fixed number of categories.

- The "Equal-width bins" method uses the same binning method as the histogram. It divides variable values into a set of equal size ranges. For example, if the minimum value of a variable is 0, the maximum value is 100, and 10 categories are specified, then the first level of the binned factor corresponds to values between 0 and 10 of the original variable, the second level of the binned factor corresponds to values between 10 and 20 of the original variable, and so on.

- "Equal-count bins" finds the appropriate break points along the range of a continuous variable so that each bin has an approximately equal number of cases, or individuals in our data. Ties can prevent the categories from being exactly equal in size.

- The "Natural breaks (from K-mean clustering)" method involves attempting to find a specified number of clusters in the data such that the members of each cluster are similar to one another, but are as different as possible from the members of other clusters. To illustrate this, consider a situation in which we have the following 10 values for a variable (1.05, 5.63, 5.71, 3.45, 3.47, 3.50, 1.10, 1.11, 5.75, 5.73), and we ask R Commander to create three groups based on the natural breaks method. In this instance 1.05, 1.10, and 1.11 would be in group 1, 3.45, 3.47, and 3.50 would be in group 2, and 5.63, 5.71, 5.73, and 5.75 would be in group 3.

Figure 3.39: Specifying Level Names Dialog

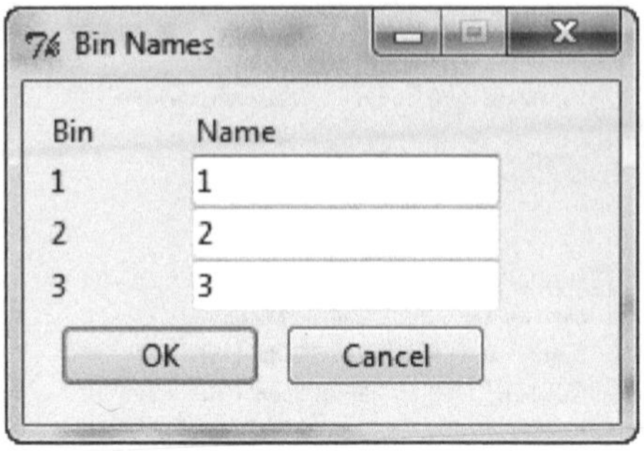

The choice of a method to use depends on the intended use of the new categorical variable. For example, in the next tutorial we will learn to use *contingency tables,* a quick, common, and useful analysis method, but only for categorical variables. A limitation of contingency tables is that we should avoid having cells in the table with very few observations, otherwise our statistical tests may not be meaningful. In converting our continuous variables to categorical, the "Equal-count bins" method is least likely to result in contingency table cells with small counts. We also want to keep the number of levels of the categorical variable small, since more levels means fewer observations in each level. We will start by creating a new categorical variable from the continuous variable by binning the spending data into roughly three equal groups (e.g., a low, medium, and high spending group).

Select SPENDING in the "Variable to bin (pick one)" field. In the "New variable name" field **enter Spend.Cat** (so you will know this new variable is the categorical version of the spending variable). Use the slider to **select 3** as the number of bins. **Select the "Specify names"** radio button among the "Level Names" choices, and **select the "Equal-count bins"** radio button among the "Binning Method" choices. Once you have done all of this press **OK**. A second dialog box (shown in Figure 3.39) will appear so that you can assign level names that are more meaningful, thus making your output easier to interpret.

9.

The data values of the original variable (SPENDING) are sorted from smallest to largest. As a result, **in the first field replace "1" with "Low,"** in the **second field replace "2" with "Medium,"** and in the **third field replace "3" with "High." Press OK** and the Spend.Cat variable will be created. To confirm the creation of the new variable, **click on the View Data Set** button, and scroll to the right to the end of the variable list where Spend.Cat will have been added. **Close the View Data window.**

Figure 3.40: Frequency Distribution Dialog

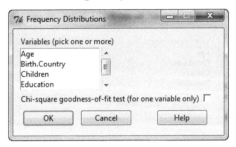

10.

As indicated above, the three groups of values for `Spend.Cat` will be of *roughly* equal sizes. We can create a frequency distribution of `Spend.Cat` to see just how "rough" equal sizes are. To do this, use the pull-down menu option **Explore and Test** → **Summarize** → **Frequency distributions...**, which will bring up the dialog box shown in Figure 3.40.

11.

In this dialog scroll down and **select `Spend.Cat` as Variable (pick one) and press OK**. The results of the frequency distribution will appear in R Commander's output window, and should look identical to Figure 3.41. The first table indicates the number of households in each bin, and the second, the percent of the total households in each bin. The frequency distribution indicates that the groups are not of exactly equal sizes, but they are close enough for our purposes.

Figure 3.41: Frequency Distribution of Binned Spending

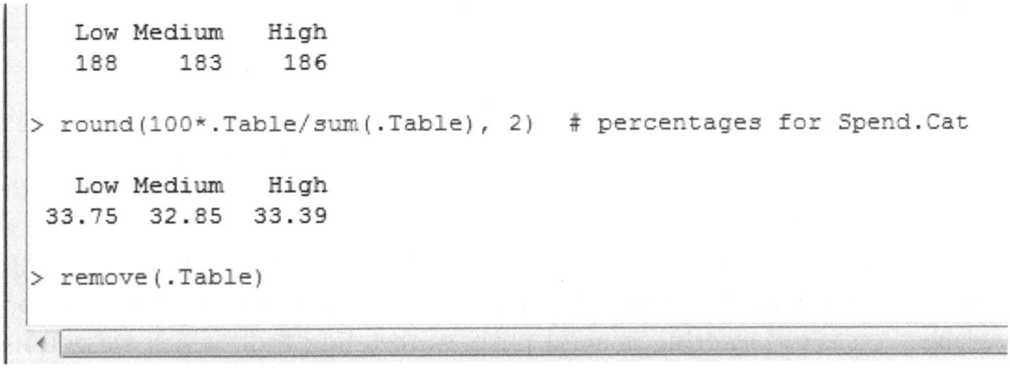

```
    Low Medium    High
    188     183     186

> round(100*.Table/sum(.Table), 2)   # percentages for Spend.Cat

    Low Medium    High
  33.75   32.85   33.39

> remove(.Table)
```

Figure 3.42: Bar Graph Dialog

12.

The final tool to be introduced in this lab allows you to take an existing categorical variable and "re-label" factor levels, which allows us not only to change the names, but to combine levels, thereby reducing the total number of levels. There are a number of reasons why we might want to do this, one of which we have already discussed. Specifically, fewer levels means more observations in each level, which may be needed for contingency table analysis of relations between variables. One variable that seems likely to be related to household spending on children's clothes is the number of children in a household. The current CHILDREN variable has four levels: "1 child," "2 children," "3 children," and "4+ children." Creating a bar graph of CHILDREN illustrates the potential problem with this variable. To create this bar graph select the pull-down menu option **Explore and Test → Visualize → Bar graph**. . . , which brings up the dialog box shown in Figure 3.42. In this dialog box **select CHILDREN and press OK** to create the bar graph shown in Figure 3.43. If you don't see it, the graphics window may be behind other windows. You can bring it to the front from the Windows taskbar, which is normally located at the bottom of your screen.

Exercise: Copy the plot into a word processing document.

13.

The bar graph reveals that there are very few households in our sample with four or more children, causing any inferences about this group's behavior in the entire population to be very imprecise and of limited usefulness for decision making. Therefore, it makes sense to combine the last two levels so that CHILDREN has only three levels (i.e., "1 child," "2 children," and "3+ children"). We have two tools that can do this. One of these tools, Recode variable, is a very powerful, general purpose recoding tool, which we will use in a later tutorial. The second tool (which can be accessed with the pull-down menu option **Data → Manipulate Variables → Relabel factor levels. . .**) is a more limited tool specifically designed for categorical variables, for relabeling

Figure 3.43: Bar Graph of the Number of Children Present

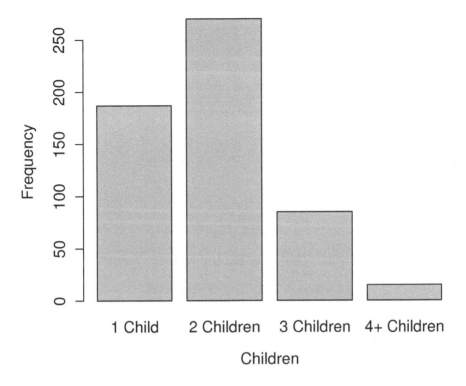

and combining factor levels. It will bring up the dialog box shown in Figure 3.44.

14.

In this dialog box select CHILDREN for the "Factor (pick one)," enter New.Children in the "Name for factor" field, and press **OK**. A second dialog box (shown in Figure 3.45) will then appear.

15.

In the **New Labels** dialog box **enter "1 Child" in the first field, "2 Children" in the second field, and "3+ Children" in both the third and fourth fields**. When you are done, your **New Labels** dialog box should appear as the one shown in Figure 3.46. When it does, press **OK** to create the new factor New.Children. View the data set and confirm the changes from the CHILDREN column to the New.Children column.

Figure 3.44: Relabel a Factor Dialog

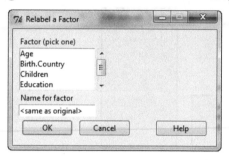

Figure 3.45: The New Factor Names Dialog

Figure 3.46: The Completed New Factor Labels Dialog

16.

Plot the graph again, select **Explore and Test** → **Visualize** → **Bar graph**......
and **select** `New.Children` for plotting.

Exercise: Copy the plot into a word processing document. Comparing with
the previous plot, approximately how much has the number of "3" families in
the `CHILDREN` variable been increased by in the "3+" case in `New.Children`?

17.

You may continue directly to the next tutorial. However, if you wish to quit
R now, first save your workspace **(File** → **Save R workspace as...** as **Tuto-
rial3.5.RData**. Then exit from R and Rcmdr as in the first tutorial: **File** →
Exit → **From Commander and R**. Do not save any other files.

3.6   Contingency Tables Tutorial

The purpose of this tutorial is to show you how to produce contingency tables.
To avoid any potential confusion, contingency tables are also called "cross-
tabulations" and "cross-tabs." They are the simplest type of *multivariate* analysis (i.e., methods for studying the relationships among multiple variables)
available and are common in market research. This tutorial will only consider
two-way (two variable) contingency tables. Interpreting three-way tables is often difficult, and four-way or higher-order tables are essentially impossible to
interpret. R Commander will also calculate a chi-square test statistic to help
you evaluate the statistical independence of the two variables under study.
You will also learn how to change the level order in a factor variable, to create
more easily interpreted and presentation-friendly contingency tables.

1.

If you are continuing directly from the last tutorial without exiting R, go to
the next step of this tutorial. If you exited R and saved the workspace file
in the last tutorial, start RCmdr by **double-clicking on the workspace
file Tutorial3.5.RData**. In R Commander, **activate the jack.jill data set
using the Data Set button.**

Figure 3.47: The Contingency Table Dialog

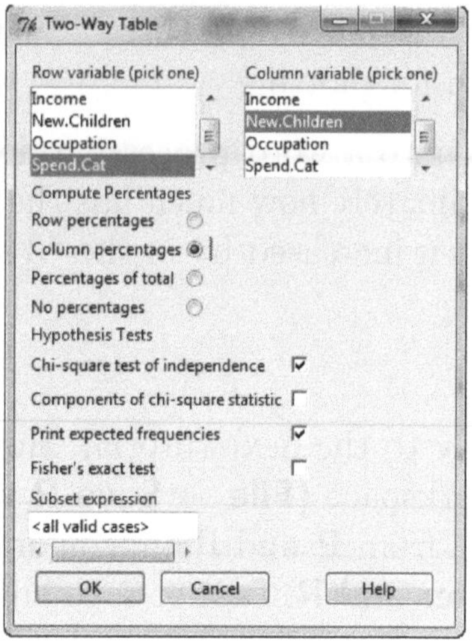

2.

Use the pull-down menu option **Explore and Test → Test → Contingency tables → Two-way table...** to bring up the dialog box shown in Figure 3.47.

3.

The first thing we will examine is the relationship between spending on children's apparel and the number of children present in the household. We should, of course, have some rough idea (i.e., "theory") about how these variables should be related! In the dialog box **select `Spend.Cat` as the "Row variable (pick one)" and `New.Children`** (the condensed version of `CHILDREN`) **as the "Column variable (pick one)" variable. Next select the "Column percentages" radio button from the "Compute Percentages"** choices. This will cause two different tables to be produced. The first table contains the raw counts (number of households) with a given level of the two factors (e.g., the number of "low" spending households with only one child present), while the second table will provide the percentage of high, medium, and low children's apparel spenders for a given number of children present, as given by each level of `New.Children`. Percentage figures help interpret a contingency table since it compensates for differences in the total number of households of different types (in this instance with differing numbers of children). By default the "Chi-square test of independence" box is checked to give us that useful statistic. **Next, check the "Print expected frequencies" option** which prints out the number of cell counts that would be *expected* in each cell *if* there was no relation between the two variables, that is, assum-

Figure 3.48: Children's Apparel Spending vs. Number of Children

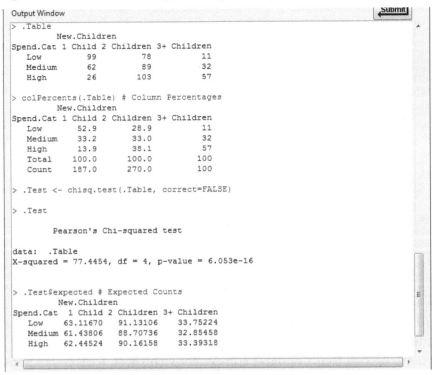

ing the variables are independent. We will not be using "Fisher's exact test," which provides an alternative hypothesis test of independence for a two-by-two contingency table (a table with two variables where each variable has only two levels). Among the "Hypothesis Tests" options, select only the chi-square test. Once you have done all this press **OK**, and the results of the table (shown in the R Commander output window) should look like Figure 3.48.

4.

The second table indicates that 52.9% of households with only one child present are in the lowest spending category, while only 13.9% are in the highest spending category. In contrast, only 11% of households with three or more children are in the lowest spending category, while 57% are in the highest spending category. For comparison, the final table shows the expected counts *if there was no relationship* between the variables. Households would then be expected to be distributed across the spending categories in exactly the same proportions as the overall distribution, which in this case is equal thirds. The actual distribution in the above tables is very different from the lower table. The extremely small p-value (6.053 times 10^{-16}) of the chi-squared test is a measure of this difference, indicating that it is essentially certain that in the

Figure 3.49: Children's Apparel Spending vs. Income

```
        Income
Spend.Cat $0-$20k $100k+ $20k-$30k $30k-$40k $40k-$50k $50k-$60k $60k-$75k $75k-$100k
   Low      31      5      23       34        33        26        25        11
   Medium   21     14      15       22        32        24        26        29
   High     17     23      10       22        21        28        23        42

> colPercents(.Table) # Column Percentages
        Income
Spend.Cat $0-$20k $100k+ $20k-$30k $30k-$40k $40k-$50k $50k-$60k $60k-$75k $75k-$100k
   Low     44.9   11.9    47.9     43.6      38.4      33.3      33.8      13.4
   Medium  30.4   33.3    31.2     28.2      37.2      30.8      35.1      35.4
   High    24.6   54.8    20.8     28.2      24.4      35.9      31.1      51.2
   Total   99.9  100.0    99.9    100.0     100.0     100.0     100.0     100.0
   Count   69.0   42.0    48.0     78.0      86.0      78.0      74.0      82.0

> .Test <- chisq.test(.Table, correct=FALSE)

> .Test

        Pearson's Chi-squared test

data:   .Table
X-squared = 46.0957, df = 14, p-value = 2.705e-05
```

whole population there is a relationship between household spending on children's apparel and the number of children present in the household (provided we have carefully taken a random sample, of course!). We have seen the nature of this relationship in the column percentages table. In sum, it is safe to conclude that households with a greater number of children present tend to spend more on children's apparel compared with those with fewer households. Of course, from a common sense view, if we did not find that this was the case, we would begin to question the validity of our sample!

Exercise: Highlight and copy the table into a word processing document.

5.

Create a second contingency table by repeating step 2, but this time use INCOME as the column variable. You can widen the RCmdr window before creating the contingency table to avoid text wrap around. Your results should look like those in Figure 3.49.

6.

One thing to notice in Figure 3.49 is that the column for the highest income group ($100k+) is out of place, falling second, rather than last, which can make interpreting and presenting the table more complicated than need be. The problem is due to the fact that, by default, R orders and prints the levels of a factor alphabetically, and "$100k+" comes before "$20k–$30k." However, it is easy to solve this problem by using the pull-down menu option **Data →
Manipulate Variables → Reorder factor levels...**, which will bring up the dialog box shown in Figure 3.50.

Figure 3.50: Reorder Factor Levels Dialog

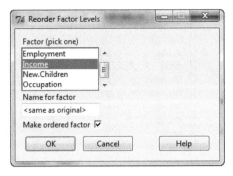

Figure 3.51: The Second Reorder Levels Dialog

7.

Select INCOME as "Factor (pick one)," and keep the factor's original name. An ordered factor is an ordinal scaled variable, while an unordered factor is a nominally scaled variable. Income is really an ordinal scaled variable, so the **"Make ordered factor" option should be checked. Finally, press OK and select Yes** when asked to Overwrite Variable. The dialog box shown in Figure 3.51 will appear.

8.

The **Reorder Levels** dialog box allows us to explicitly reorder the factor levels. The "$0- $20k" level is correctly placed in the first position, but the remaining levels are out of place. For instance, the "$100k+" level should be in the eighth (not second) position, and the "$20k–$30k" level should be in the second (not third) position. Reorder the factor the way it should be. When you are done, this dialog box should look like the one shown in Figure 3.52. When it does, press **OK**.

Figure 3.52: The Completed Reorder Factor Level Dialog

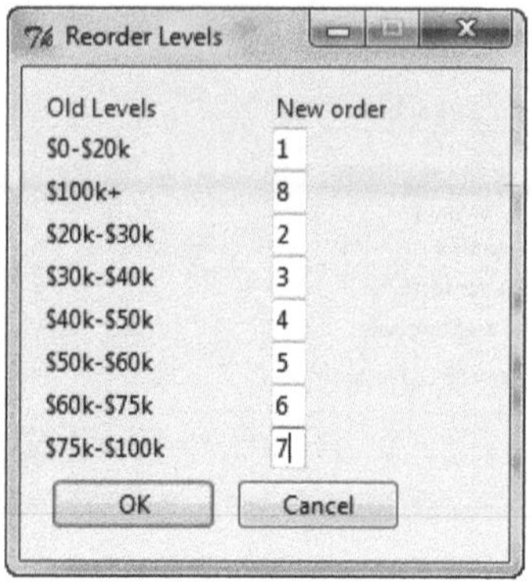

9.

Re-create the `Spend.Cat` and `INCOME` contingency table using the reordered `INCOME` variable and examine the relationship between children's apparel spending and household income.

Exercise: Highlight and copy the table, and paste it into a word processing document.

10.

Next we explore three additional relations with Spending: (1) `Spend.Cat` and `EDUCATION`; (2) `Spend.Cat` and `AGE`; and (3) `Spend.Cat` and `BIRTHCNTRY`. First explore these three variables individually **using frequency distributions.** Note that there are levels with very small counts, which will not be great for our tables. Also (if you inspect the help file for this data) you will see that these variables are specifically for the *female head of the household*. In the real world there are households with children and no mother. Therefore there are no data for these variables in these households. The frequency distribution shows that this occurs 9–10 times. That is few enough that we can quickly remove those households individually. Click on the **View data set** button to see which rows the offending variable occurs in. For example the very first household has no female head. As long as we have set the variable HH.ID as the case name, you will see that this record is named "7650," which R interprets as a character string rather than a number because of the quotes surrounding it. Select **DATA→ Clean→ Remove Selected Records** to bring up a dialog box. You can delete the record by its row position, in this case 1, or by its name, in this case "7650," with quotes. Enter "7650" in quotes. Leave the data set name the same and Click **OK.** The first row of the data will now disappear. **Repeat the exercise for the remaining households that have no female head.**

Now with your cleaner data, create your contingency tables.

Exercise: Copy and paste the three tables into the word processing document.

11.

Exercise: Based on the five contingency tables you have done, the chi-square statistic, and the expected counts (to assess the validity of the statistic) which demographic and socioeconomic factors have the largest impact on household spending for children's apparel? Which appear to have no impact?

12.

Since we are now finished with the data, you can exit without saving the workspace file. However, you may wish to save your output file for future reference using **File → Save output as. . .** since it will contain the contingency tables you have generated.

Chapter 4

Multiple Linear Regression

Linear regression is the oldest and by far the most commonly used predictive modeling method. The biologist Sir Francis Galton, a cousin of Charles Darwin, is credited with its introduction in 1877 (Bulmer, 2003). He was actually less interested in prediction per se than he was in studying how traits were passed down from one generation to the next. He plotted characteristics of one generation against characteristics of the next, and noted that he could draw a straight line through the plot. Sweet peas that were heavier than average tended to produce offspring sweet peas that were heavier, tall fathers tended to have tall sons, and smart parents tended to have smart kids. However, what initially intrigued him was that the offspring of individuals with more extreme characteristics, that is, characteristics that are far from the mean of the population, tended to be closer to the mean than their parents. He referred to this as "regression to the mean," and since then we have been stuck with the rather uninformative name "regression" for the method.[1]

Since that time, a massive amount of statistical machinery has been developed to refine and extend Galton's seminal ideas. It is worthwhile to note that if we only have two variables—a single predictor variable and a target variable—this statistical machinery is overkill. We can do very nicely by using Galton's simple method of plotting the data. The reason for the algebra and associated number crunching is simply that there isn't much of interest that we can predict with only one predictor variable, and as soon as we get more than one, plotting goes quickly from impractical to impossible. With 3D plotting software we can visualize two predictors and one target, but that is the limit of conventional plotting in the three-dimensional world we live in. For anything else we need algebra, which is not limited to three dimensions.

The problem faced by those early statisticians was how to do with numbers what Galton did with his eyeball and his plots. Once that was figured out for one predictor and one target, it was relatively easy to generalize the approach to multiple predictor variables. The resulting method was called multiple linear

---

[1]"Regression to the mean" is a purely statistical effect and has nothing to do with genetics. In sports, an example of this is the so-called "sophomore slump." A player who has an outstanding rookie season must credit some part of the success to "luck" (really random chance). That luck can't be counted on in the next season, and so the player, now a sophomore, is likely not to do as well.

Table 4.1: Variable Roles

Statisticians say...	Data Miners say...	Meaning...
Independent Variables	Predictor Variables	Things you can use as inputs to predict an outcome
Dependent Variable	Target Variable	The outcome you want to predict with the inputs

regression, or more conveniently, *multiple regression*. While Galton understood the idea of multiple regression, he was unable to formalize the algebra. It was left to Karl Pearson, of Pearson Correlation Coefficient fame, and a colleague of Galton's, to master this tricky problem (Bulmer, 2003).

In our exposition, we will follow this historical logic, by first demonstrating the key concepts of regression using Galton's method of plotting one predictor against the target variable, and then translating those insights into the associated algebra. Most of the calculations necessary to generate numbers like coefficients and p-values will remain behind the scenes in our discussion, since today we can rely on numerically fluent computers to handle those details for us.

Once the foundations have been established, we will describe two useful extensions:

1. Incorporating categorical predictor variables

2. Handling nonlinear relationships with a linear model

We will also review the meaning and use of the coefficients, p-values, and R^2 that appear in the output of regression software.

4.1   Jargon Clarification

To start, we will clarify a bit of jargon, since statistical methods and data mining both developed in a number of different fields, with the different fields using different terms to describe the same concept. As an example, the terms describing the basic roles of variables differ across fields, as Table 4.1 illustrates.

We present the usual cautionary note that there is an implicit notion of causality here—the predictors are considered to cause the target outcome. However, causality is never proven by this method alone when applied to actual data. Only data from controlled experiments can allow us to conclude that a predictor causes the target. Regression on data that is not from controlled experiments can only indicate that changes in a target and a predictor variable occur together. Like chickens and eggs.

4.2   Graphical and Algebraic Representation of the Single Predictor Problem

Speaking of eggs, imagine a problem facing a supermarket category manager in southern California, who puts in orders for one week of eggs every Monday, and would like to predict the current week's volume of egg sales to know how much to order. The manager knows today's egg price per dozen, $1.05, and she has about two years of historical data available:

1. Target variable: number of cases of eggs sold each week

2. Predictor variable: weekly egg prices

Let's start by looking at the raw data. Some of these data are shown in Table 4.3. Even with only two variables, extracting useful information from a table of numbers is extremely difficult.

Let's try Galton's trick of plotting the numbers and see if things become more transparent. Figure 4.1 shows the scatterplot, and it does indeed look more informative. We can use this scatterplot to look up historical sales. Before reading on, try it—answer these two questions from the graph: If egg prices this Monday are at $1.05, what would you expect the weekly sales of eggs to be? What would you expect sales to be at a price of $0.80?

In coming up with sales figures, you have just used a predictive model. Congratulations! You probably came up with an estimate of around 103,000, plus or minus 5,000 for the first, and 95,000, plus or minus 15,000 for the second. Of course, you very likely also want to add some cautionary notes to those sales forecasts if it is to be used as a prediction for the upcoming week. Again, stop for a moment and make a mental list of some of those cautionary notes.

We will mention three things here.

Table 4.3: Egg Prices and Sales

Cases Sold/Week	Egg.Prices
96343	90.42
96345	89.33
96928	89.89
93519	90.71
99032	85.99
91539	91.83
89969	87.29
90859	96.36
99697	99.71
88350	99.38
100383	97.53
94415	100.77
91813	98.00
100466	99.89
96783	101.26
91008	101.26
100324	100.28
106628	100.69
98892	105.16
98252	109.55

Figure 4.1: Weekly Eggs Sales and Prices in Southern California

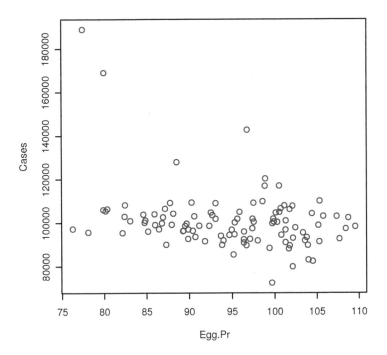

1. The relationship between price and sales for the coming week may not be the same as the past relationship. This problem always exists when using historical data as a basis for prediction, and the analyst needs to apply common sense and experience to judge if the past relations are likely to be relevant this week. Obviously, if the analyst is also a manager with experience in the industry, he or she will be in a good position to make that judgment.

2. The scatterplot indicates a range of possibilities for the prediction. Take a look at your sales prediction for $0.80 and $1.05, and this time include a range of possible values. A good way to think of the range is to think to yourself "what is the range of values that I would be 95% sure that the sales this week will be within?" Before reading on, look at the scatterplot and try this for $0.80 and $1.05.

3. Notice that within the scatter of the data points, egg prices don't seem to have too much effect on sales. The cluster of data points is rather flat. Granted, our estimate of sales is lower for higher prices, which is reasonable, but the range of sales at any price is large enough that we would not be wildly confident of lower sales with higher price in any specific case.

That was fairly easy, intuitive, and only took elementary school math skills. However, since our ultimate purpose is to tackle problems with many predictor variables, we will next consider how to do the same analysis with algebra, so that we have a framework that can be generalized to beyond the two-dimensional limit of scatterplots. In short, we want an equation into which we can plug egg prices, do a bit of arithmetic, and calculate sales. The usual approach to this problem is in two steps. First, we narrow down the range of all possible equations to a set with a particular functional form or structure. The simplest structure is between two variables is one of strict proportionality, with some constant of proportionality. We can write the set of all possible equations of this form as

$$\text{Sales} = b \times \text{prices},$$

where b is the constant of proportionality. All possible equations of this form are given by all possible values of b. The second step is to select the particular numeric value for b that describes the relationship in our particular data set.[2] This step is called *calibration*, or *estimation*, or *parameterization*.

This simple structure is very restrictive. The biggest restriction is that when prices are zero, sales must be zero. As such, it is not a very good model for this, or many other, situations. We can make it more flexible, and applicable to more situations, by adding another parameter, a, which will be the level of sales when egg prices are zero.

$$\text{Sales} = a + b \times \text{prices}$$

With all possible combinations of values of a and b, we now have a much larger set of possible relations between prices and sales. Again, we would use the data to *calibrate* (or estimate) the equation (i.e., to select specific values of a and b). Of course, once calibrated and we have specific values for a and b, we could plug in any value for price (such as \$1.05) and calculate the predicted value for sales. It turns out that if we plot the combinations of sales and price given by this equation for specific values of a and b, they fall on a straight line. Hence, we call this a linear model, and a linear model is the structural form assumed by linear regression. The new parameter a is called the *intercept*, because the line defined by the equation crosses, or intercepts, the sales axis at the value of a. The parameter b is the *slope* of the line. The linear model is of course

---

[2]By a "particular numeric value" we mean a number, like 2.45 or 763,457, rather than a symbol like b.

still restrictive, and many other structures in the relationships between two variables occur in the real world.[3]

If, as was the case with Galton's sweet peas, we start with the scatterplot and draw a straight line through the data, and it looks like that line approximates the data well, we could measure the slope and intercept of the line on the graph. We could substitute those values into the equation for the linear model, and now have an algebraic version of the scatterplot. Given prices, we can find sales. The question remains, though, how do we use the data, *without* making a scatterplot (since we can't do that when we have many predictors), to determine particular values for the parameters *a* and *b*? One obvious answer is to try and duplicate what we do by eye when we draw a straight line through a bunch of scattered points, which is to try and put the line through the middle of the points. We could duplicate this numerically by adding up the distances of all the points on the scatterplot from a trial line, and then adjusting the line so that this total distance is minimized. Other variations on this rule could be imagined. The particular rule in linear regression is to add up the the square of the distances of the points from the line in the *y*-direction, and to minimize that.[4] It turns out that surprisingly simple formulas can be found for *a* and *b*, given a set of data points, using this rule. We will not, however, go into those details. We will simply use R to take our data and calculate *a* and *b* using those formulas. The result is that *a*, the intercept or the expected sales level when the price is zero, is 153414 cases of eggs. The coefficient of prices, or the slope, is −554. The regression equation, which we can now use for prediction, is

$$\text{Sales} = 153414 - 554 \times \text{prices}$$

Note that the slope is negative, which means that as prices increase, sales decrease in this predictive model, as we suggested when we did our eyeball predictions. By the way, we should take that to mean that our numerical methodology seems to work because it duplicates our eyeballing, not that our eyeballing is confirmed by the method. Let's plot the values of sales and prices that satisfy this equation on top of our scatterplot, as in Figure 4.2.

Interestingly, this may not be the line you would have drawn by hand. If you agreed with the eyeball prediction estimates above, you likely would have

---

[3]Such as diminishing returns, increasing returns, U-shaped, and so on. We typically divide the universe into linear relationships and everything else, called nonlinear relationships. This is rather like dividing the zoological universe into pachydermic and non-pachydermic animals.

[4]To see an interactive animation that allows you to manually adjust the slope and intercept of a line to adjust the squared *y*-distance to a set of points, see http://www.dynamicgeometry.com/javasketchpad/gallery/pages/least_squares.php.

Figure 4.2: The Regression Line and Scatterplot

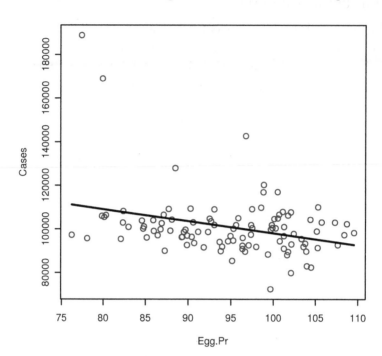

drawn a flatter line. Again, we should take that as a problem with the method, not with our eyeballing, since we are trying to find a numerical way to duplicate what we have easily done by eye graphically. We will come back to this in a minute. First, let's recall that when we used the scatterplot for sales predictions, we were also able to provide some sense of the possible range of the prediction. The formula, and the associated straight line, does not provide this useful information. Fortunately, the statistical machinery has also been developed for assessing this range, or variability. The machinery translates the scatter in the data to uncertainty in predictions of sales. Once again, we will not delve into the details of the calculations, but simply make the observation that if we can do this by eye, it should not be too surprising that we can find ways to do it with numbers. Just as we can show the calculated regression equation as a straight line to get a visual interpretation, we can plot the calculated uncertainty associated with the prediction line as error limits around that line (Figure 4.3) to provide a visual interpretation of the prediction error.

Although not quite the precise technical description of confidence limits, we can think of the prediction error shown here as a 95% chance that the actual realized case sales will fall between the dotted lines. This range is probably quite close to the range you guessed during the eyeballing exercise.

Figure 4.3: 95% Confidence Limits of the Regression Prediction

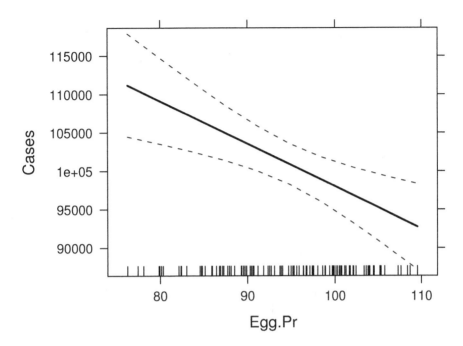

4.2.1   The Probability of a Relationship between the Variables

In many regression applications it seems that the main question of interest to the analyst is the rather weak one of whether the predictor variable is likely to be related to the target variable at all, given the data available. A slope coefficient of zero, corresponding to a horizontal regression line, means that there is no relation between the variables. That is, knowing the value of the predictor variable cannot refine your estimate of the target variable. Your best guess is always the mean of the target regardless of the value of the predictor. The scatterplot in Figure 4.1 suggests that we are close to the zero slope case for the relation between egg prices and sales. In Figure 4.3, however, the slope is quite steep. In fact, if we were to twist the line to horizontal some of it would have to be outside the 95% confidence limits for prediction, which suggests that it is highly unlikely that the true slope could be zero. This suggests asking a more refined question: "What is the probability that the true slope is actually zero?"

To answer this we need to know the range of uncertainty in the value of the slope coefficient. This uncertainty calculation again uses the scatter of the data. The more scattered the data, the greater the possible range of the coef-

Table 4.5: Regression Output for the Eggs Data

```
Coefficients:
                Estimate  Std. Error  t value  Pr(t)
(Intercept)     153414.5     15992.9    9.593  5.96e-16 ***
Egg.Pr            -553.9       168.2   -3.293   0.00136 **
Signif. codes:0 *** 0.001 ** 0.01 * 0.05 . 0.1  1
```

ficients. The slope coefficient uncertainty can be expressed in a variety of ways, such as variance or its square root, the standard deviation, or as confidence intervals. The calculated standard deviation of the estimate of the slope coefficient for this regression is 168. With our calculated slope of -554 (which we can now call the expected or most likely slope), a slope of -486 is one standard deviation closer to zero ($-554 + 168 = -486$). A slope of zero would be 3.29 standard deviations from the most likely slope. ($-554 / 168 = 3.29$). Since we know the probability distribution of the slope that we have calculated we can say something specific about the probability that there is no relation (i.e., a zero slope). This procedure is known as "hypothesis testing."[5] The way the calculations are done, however, gives a slightly different interpretation. The actual refined question becomes IF the true slope were zero, how likely is that we would have observed a sample of data that gave a slope equal to or *less* (in this case) *than* the calculated mean slope of -544. From the t-distribution, this probability is 0.00136, or highly unlikely. While not technically correct, we can also think of this as indicating that it is highly unlikely that the true slope is zero, or highly like that there is indeed a relationship. All of the above information is contained in the regression printout in the row starting with "Egg.Pr," shown in Table 4.5. It is very worthwhile to work across this row and ensure that you understand the meanings of each of the numbers by relating then back to the preceding discussion and plots.

The stars beside the probability are a bit of a holdover from the days when distribution tables were used, and exact probability values were not calculated. They indicate the range in which the probability falls, and are interpreted in the last line as significance codes. Two stars, for example, means that the probability is between 0.01 and 0.001, and the categories are referred to as

---

[5]Or, more correctly, the people who programmed the software know the distribution. It was derived by William Gosset, a statistician and chemist employed by the Guinness brewery in Dublin from 1899 until his death in 1937. The story goes that Arthur Guinness was one of the earliest businessmen to recognize that skill in quantitative methods conferred a powerful competitive advantage, and did not want competitors to know that he was using them. He therefore prohibited employees from publishing their results. Hence, when Gossett published his results in 1908, he used the pen name "Student" (Pearson 1990). It was Fisher who labeled the statistic "Student's t." The rest is history.

statistical significance levels. From this, it should be obvious that calling a coefficient significant, or not significant, is a bit arbitrary, as it depends on what level of significance you are willing to tolerate. Which amounts to how often are you willing to be wrong when you label your result "significant." We find it more useful to report the p-value and leave the reader to judge whether that qualifies as "significant."

4.2.2 Outliers

At this point, return to Figure 4.2. Note that the calculated regression line slices diagonally through the main cluster of data points. If you were predicting sales simply by eyeballing the scatterplot, you might well make your predictions at low and high prices closer to the center of the cluster of points than this line would suggest. The difference, of course, is that the calculations that give the line are strongly influenced by a few very high sales figures on the left side of the plot. This creates a problem: Should we really use these outliers to adjust our sales estimates upwards at low prices? Or should we treat them as anomalous values and ignore them? This is not an easy question to answer, but we should first try to figure out why the anomalous values occur. They could be just a typographical error during data entry; they could be true but inexplicable values; or they could be true, and we might be able to figure out what caused them and use that information for future predictions. If we cannot come up with a reason for the outlier, it is usually best to drop it from the data set and re-estimate the regression, which will give us a prediction formula that is more like an "eyeballing" prediction that chose to ignore these large sales values. We will pursue the course of finding a reason for these outliers next.

4.3 Multiple Regression

For our eggs data, we also have other potential predictor variables available, and by exploring these variables a bit, we find that the outliers occur around Easter. Common sense says that means that we can make a prediction that starts with a price effect on sales which do not occur at Easter (in the eyeballing case, that means ignoring the outliers), and then, if our Monday eggs order is around Easter, make an adjustment to that prediction. Once we've identified when the outliers occurred, this is easy enough. To do the same numerically, we need to add at least one more predictor variable to the analysis, and we are now into the wonderful world of *multiple* regression. Graphical

Figure 4.4: The Data View for the Eggs Data Set

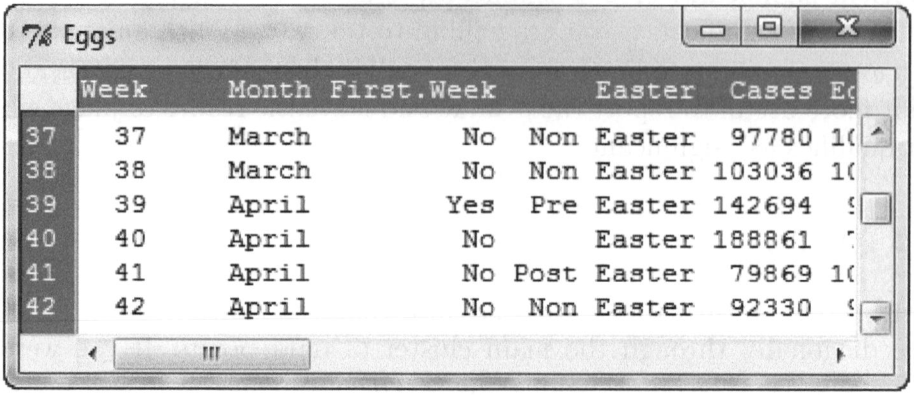

	Week	Month	First.Week		Easter	Cases	Eg
37	37	March	No	Non	Easter	97780	1(
38	38	March	No	Non	Easter	103036	1(
39	39	April	Yes	Pre	Easter	142694	(
40	40	April	No		Easter	188861	
41	41	April	No	Post	Easter	79869	1(
42	42	April	No	Non	Easter	92330	(

plots rapidly become overwhelmed, but the algebra we have developed can be extended quite easily.

4.3.1   Categorical Predictors

As long as our predictors are continuous (interval or ratio scale) tossing more of them into the regression algorithm and interpreting the outcome, is fairly straightforward.[6] "Easter," however, is a categorical variable. We can write

$$\text{Sales} = a + b \times \text{Price} + c \times \text{Easter}$$

and understand that we replace price with a number, like \$1.05. But what do we replace Easter with? Furthermore, if we look at the actual data, we see that "Price" is a variable name at the top of a column that contains numeric values, but in the "Easter" column the values are "Non Easter, Pre Easter, Post Easter, and Easter" (Figure 4.4).

We cannot add, subtract, multiply, etc. "Non Easter," and so can't just throw the values into the number cruncher. In fact the "Easter" variable is nominal, and as we have seen in a previous chapter, you cannot do arithmetic on nominal variables. The way around this is to create new variables based on the Easter variable, which will indicate with a zero or one what week we are in. Our first indicator variable will indicate whether or not it is Easter, and we will call it "IndEaster," and assign it the following values:

$$\text{IndEaster} = 1 \text{ when Easter (the variable)} = \text{Easter (the value)},$$

---

[6]There is one major pitfall, namely correlated predictors, or *multicollinearity*, which we will not discuss in this chapter.

IndEaster $= 0$ for all other values of Easter

"IndEaster" contains numbers—numbers that our software can crunch—whose value depends on "Easter." The prediction formula now becomes

$$\text{Sales} = a + b \times \text{Price} + c \times \text{IndEaster}.$$

Note what happens once we have the coefficients estimated and go on to use this formula for prediction. When the week we are predicting is not Easter, the value of IndEaster is zero, and the last term ($c \times \text{IndEaster}$) is zero. Sales will depend only on price. But if it is Easter week when we wish to predict sales, IndEaster is 1, and the last term takes the value we have estimated for the coefficient c. This means that the value of c is the sales boost that occurs during Easter. As a bonus, because the Easter outlier data points are now taken care of by IndEaster, they no longer affect the other coefficients (they don't pull the price line up at the low prices) so that the coefficient of price will be smaller, and the prediction regression line flatter.

We can add additional indicator variables for Pre Easter and Post Easter. Note that we have four values for the categorical variable, but we can only create three indicator variables. The coefficients are interpreted as the change in sales relative to the value for which we have not created an indicator variable, namely Non Easter. This is a general rule: If a categorical variable has n values, create $n - 1$ indicator variables, and interpret their coefficients as the change relative to the nth variable.

When we estimate our new multiple regression equation, we get the following:

```
Coefficients:
```

	Estimate	Pr(t)	
(Intercept)	115387.19	2e-16	***
Egg.Pr	-170.15	0.0813	
Easter[T.Pre Easter]	32728.55	1.94e-08	***
Easter[T.Easter]	76946.67	2e-16	***
Easter[T.Post Easter]	-22096.43	8.25e-05	***

The coefficient estimates give the *sales prediction formula*

$$\text{Sales} = 115387 - 170 \times \text{Price} + 76946 \times \text{Ind Easter}$$
$$+ 32728 \times \text{Ind Pre Easter} - 22096 \times \text{Ind Post Easter}$$

We will note two things. First, if we are predicting egg sales during the Pre Easter week, we need to add 32,728 cases to the price-only prediction. If it is Easter, we add 77,000 cases, and if it is the week after Easter, everybody is sick of eggs, and we need to subtract 22,000 cases from our price-only prediction.

Second, now that we are explicitly considering the Easter season (some jargon: we say "controlling for Easter"), the slope of the price line is much shallower at -170. This is about one standard deviation away from zero. If the true value were zero, the chance of getting these data is 0.08 (about one in 12), which might cause some concern to the analyst who is asking, "Is there any relation at all between egg prices and egg sales?" This analyst might also say (more jargon) that the price coefficient "is significant at 0.1 level" since 0.08 is less than 0.1, but more than 0.05.

R Commander will also give you plots of the effect of one variable at a time, holding other variables constant, called *effect plots*. Figure 4.5 shows the effect plot for price with the new coefficients, as well as the 95% confidence intervals. Note that now the line could be horizontal or even positive and predictions from it would be within the confidence limits. Now that we have controlled for the Easter effect, we are much less certain that there really is a price effect on egg sales.

4.3.2   Nonlinear Relationships and Variable Transformations

At the beginning of this chapter, we said that the first step in translating our graph and eyeballing method to algebra was to decide on the model structure, or the functional form of the equation we will use. We then chose the linear model, and have spent all of our time since then dealing with the second step in the translation, namely *calibrating* the linear model. As noted in footnote 3, the world is awfully nonlinear. Nevertheless, the algebra of the one-predictor linear model seems really useful in extending the analysis to multiple predictors. We will give two responses to this little conundrum.

First, a linear model may be good enough. This is a judgment call based on how nonlinear the relationship is between each predictor variable and the target. We can use graphical visualization techniques to explore the relationships and decide whether any curves and twists we see in the data are big enough to justify the extra work of nonlinear modeling. In essence, we have to ask if approximating those curves and twists with a straight line will cause our prediction to be bad enough to cost us more from bad predictions than than the cost and effort of nonlinear modeling.

Second, we may have a relation between the variables that can be converted to a linear relationship by transforming the variables. A common nonlinear

Figure 4.5: Egg Price Effect Plot When Controlling for Easter

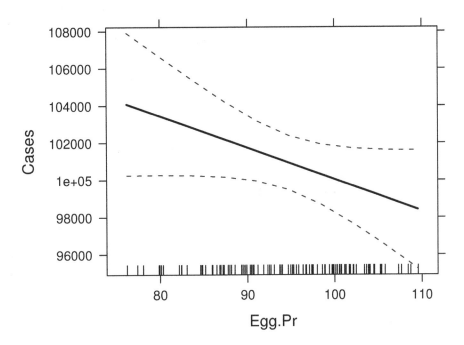

relationship between a predictor and a target is of the diminishing-returns variety, as shown in the stylized example in Figure 4.6.

If our graphical explorations of the data showed something like this, we would know that a straight line through those points would be a fair, but not great, approximation of the data. However, if we were to take the natural logarithm of the predictor variable (the one on the x-axis) and re-plot the results, the data points come closer to a straight line (Figure 4.7). Now, if we use a linear model on this transformed data, we can get a better fit, and better predictions.

Besides simply exploring relationships in the data graphically, we sometimes have theory or experience that tells us likely model structure. One common example of a nonlinear relationship that can be easily converted to linear is the power function model. For our eggs example, a power function model of egg sales as a function of egg prices, cereal prices, and chicken prices looks like

$$\text{Sale} = a(\text{EggPr})^b(\text{CerealPr})^c(\text{ChickenPr})^d,$$

where a, b, c, and d are the parameters we wish to estimate. We transform

Figure 4.6: A Diminishing Returns (Concave) Relationship

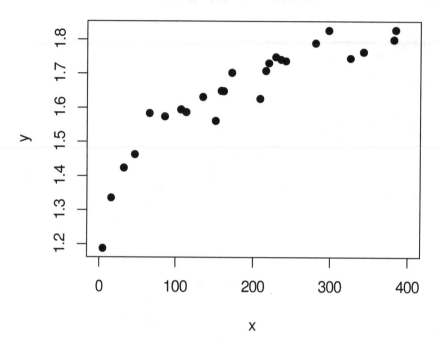

Figure 4.7: The Relationship after Logarithmic Transformation

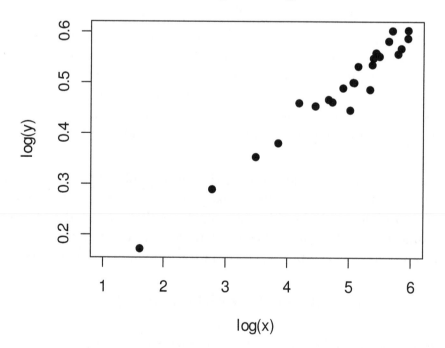

this model into a form we can apply linear regression to by taking the natural logarithm of both sides,[7] resulting in the equation

$$\log(\text{Sales}) = \log(a) + b[\log(\text{eggPr})] + c[\log(\text{CerealPr})] + d[\log(\text{ChickenPr})].$$

This is now a linear model with target variable being log(Sales) rather than Sales, and the predictor variables being the terms inside the square brackets. When we run a multiple linear regression with these logged variables, the resulting estimated intercept is $\log(a)$. To find a, we simply take the inverse, $\exp(\text{intercept}) = \exp(\log(a)) = a$.[8] Similarly, the exponents in the power function formulation, b, c, and d, are the estimated coefficients.

4.3.3 Too Many Predictor Variables: Overfitting and Adjusted R²

The last point we will touch on is the question of how good the formula is at approximating the data. The measure which we use to provide some initial sense of this called the *coefficient of determination*, or simply "r squared" (written as R^2) which takes a value between zero and one. This is one of many *measures of model fit* used in statistics, and typically the first one that people encounter (we will encounter others, for other models, as we progress through the book). Intuitively, the closer the data points lie to the regression line, the closer the R^2 value is to one.

When we have many possible variables to use, as is common in data mining, we would like to have some idea of which ones we should use and which ones we should not include. One might think that as long as the fit keeps improving as you add variables, you might as well keep adding variables. An interesting phenomenon, however, is that the more variables included in the regression, the higher the R^2 value is likely to be, *even if there is absolutely no relation between the additional variables and the target variable* (Judge et al., 1982, p. 601). The likely improvement in fit will occur because each new variable will be able to randomly account for some of the variation in the target variable. That means that R^2 by itself will not work as a means of deciding how many variables to include. One way to address this problem is to reduce the value of R^2 by a small amount every time an additional variable is added. Ideally, this amount should be roughly the amount we would expect the value of R^2 to

---

[7]R, like most statistical software, defines the function log() to be the natural logarithm with base e, a constant with a value of approximately 2.71828, while the function that takes the base 10 logarithm (the logarithm you most commonly worked with in high school math classes) is log10(). In the book, we adopt R's convention for referring to the natural logarithm function of x as $\log(x)$.

[8]exp(x) is notation for e raised to the power of x.

increase due to chance alone. The more variables included in the regression, the more we subtract from the calculated value of R^2. Statistical programs will all do this calculation, and the output is called *adjusted R^2*. Adding variables will never decrease the R^2, and will typically increase it simply by the small bit of fit improvement that a purely random relation will generate, but may decrease the *adjusted* R^2. If the increase in R^2 is very small, the adjusted R^2 will decrease with the addition of variables. In essence, the adjustment attempts to compensate for purely random fit improvements. An automatic way to decide on the number of variables, therefore, is to maximize adjusted R^2. This helps prevent fitting pure random noise, which we call *overfitting*. While maximizing adjusted R^2 is one approach, it may not be the best approach. There is both a technical and practical reason for this. From a technical perspective it has been shown that adjusted R^2 does not actually adjust enough as additional variables are added to a model, so maximizing adjusted R^2 is not a guarantee against overfitting (Amemiya, 1985; pp. 49–51), but it is a lot better than relying on R^2 alone. The practical reason is that we also need to be concerned about whether the variables included in a model make sense from a managerial perspective, rather than including them purely because of their impact on adjusted R^2. We will have more, much more, to say about overfitting in later chapters.

4.4   Summary

1. If you have learned the terms Independent and Dependent, start getting used to **Predictor and Target.**

2. **Graphical and Algebraic Representation**, or who needs algebra? You do, but only if there is more than one thing affecting the target you are trying to predict.

3. The p-values give you an indication of how probable it is that there is no relationship between the predictor and target variable. This is called **statistical significance. It does not tell you anything about the size or importance of the effect.**

4. Problems and Solutions:

   • There is a lot of stuff printed out in the regression output.

      – The most important pieces of information for now are the coefficient values, their p-values, and adjusted R^2.

- Help! I've got a categorical predictor variable!

  – RELAX. We'll use **indicator variables**, aka dummy variables.

- The real world is awfully **nonlinear.**

  – Transform your variables, on the basis of theory, your experience, or explorations of the data.

4.5   Data Visualization and Linear Regression Tutorial

This tutorial has two purposes. First, it demonstrates some of the data visualization methods provided by R for exploring the relationships between a *continuous* (interval or ratio scaled) *target* (dependent) variable and *continuous* or *categorical predictor* (independent) variables. In addition, these plots can suggest the shape of the relationship, such as linear or diminishing returns. The second purpose of this tutorial is to provide an opportunity for you to become familiar with estimating linear regression models, including handling categorical predictors, and one common type of nonlinear model structure.

1.

For this tutorial we will be using the Eggs data set from the BCA data library, which can be accessed with **Data → Get From → R Package → Read data set from an attached package...** pull-down menu option (described in more detail in previous tutorials). The Eggs data set provides information on weekly retail sales of eggs in southern California over a two-year period. Our objective is to determine what factors influence egg sales so that we (in the role of a retailer) can make better decisions regarding our pricing, ordering, and inventory decisions for eggs. You can view the R help file describing this data set using the pull-down menu option **Data → Clean → Help on active data set (if available)....**

2.

Scatterplots

We are going to begin by graphically examining the relationship between egg prices and egg sales using a scatterplot. To do this use the pull-down menu option **Explore and Test → Visualize →Scatterplot...**, which will bring up the dialog box shown in Figure 4.8. Since a scatterplot is used to examine continuous (ratio and interval) variables, those variables in the data are

Figure 4.8: Scatterplot Dialog

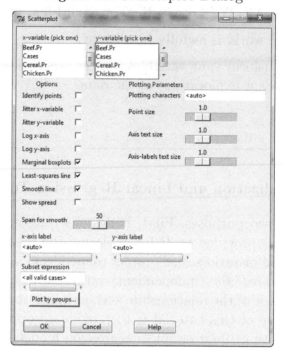

available in the dialog. **Select Egg.Pr (egg prices), the predictor, as the *x*-variable, and Cases (weekly egg case sales), the target variable, as the *y*-variable**. In addition, **check "Marginal boxplots," "Least-squares line," and "Smooth Line" boxes**. The "Marginal boxplots" option places a box plot of each of the two variables along each variable's axis, which gives a sense of the distribution of each of the variables in the plot. The "Least-squares line" and "Smooth Line" produce a plot of the fitted relationship between the two variables. The "Least-squares line" is the calibrated simple ("simple" means only one predictor, in contrast to "multiple") linear regression model, where Cases is the target (or "dependent") variable and Egg.Pr is the predictor (or "independent") variable. "Smooth Line" captures potential non-linear relationships in the data (the specifics of the method used to estimate the "Smooth Line" are beyond the scope of this book).[9]

3.

Once you have made the needed selections, **press OK**. A graphics window with the scatterplot will appear (Figure 4.9). An examination of the scatterplot points indicates that there appears to be a slight negative relationship

---

[9]For the curious, it is created using a "local regression" model known as *loess* (Cleveland and Devlin, 1988).

Figure 4.9: The Scatterplot of Eggs Sales and Prices

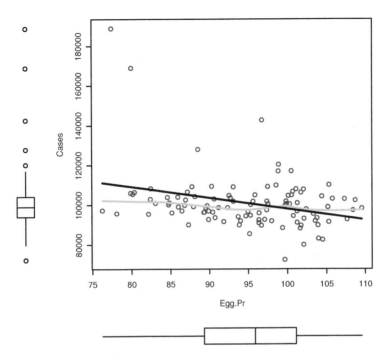

between egg sales and egg prices, and more noticeable negative slope on the least-squares (dotted) line. It also appears that the outliers are driving this negative slope. The negative relationship is less clear based on the smoothed (solid) line. The other use of the smoothed line is to look for evidence of non-linear relations between the two variables. If we see any simple, systematic curvature across the whole plot, we would likely try some nonlinear transformations of the variables, and then rerun the regression to improve the regression fit and capture the nonlinearity. In this case, there is very little evidence of any systematic curvatures, so we would be unlikely to attempt any transformations. Very shortly we will discuss how to interpret the box plots along the axes. First, let's see if we can find a reason for the outliers. If we cannot explain them we may want to drop them out of the analysis.

4.

Line Graphs

Another way to explore and visualize the relation between two variables is to create what is called a "line graph" in R Commander, between `Cases` and `Week`. Line graphs make sense *if the data is sorted* by one of the variables used in the analysis, which turns out to be true for the variable `Week` in this instance. This is often particularly useful where the sorted variable represents

Figure 4.10: Line Plot Dialog

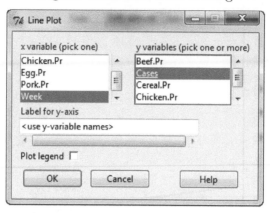

time. To create this line graph use the pull-down menu option **Explore and Test → Visualize→ Line graph...**, which will cause the dialog box shown in Figure 4.10 to appear.

5.

In this dialog box select Week as the x variable, Cases as the y variable, and press **OK** to produce the plot shown in Figure 4.11. We can immediately see the same pattern of outliers appear about 52 weeks, or one year apart, and therefore represent something seasonal happening with egg sales. This then suggests that we should look back at the data to see if we can tell what the seasonal effect is.

6.

View the data (click on the View Data Button) to inspect what else is happening during the outlier weeks 40 and 91. Happily for us, the Eggs data set includes the variable Easter, a nominally scaled categorical variable (or "factor") that captures the "Easter effect" by identifying Easter week, the week before Easter, and the week after Easter. The two very pronounced spikes in weeks 40 and 91 shown in Figure 4.11 are a result of the Easter holiday, where there is a tradition of coloring hard boiled eggs for children on Easter. Comparing the figure with these values in the data reveals that sales of eggs start a steep climb the week prior to Easter Sunday, peak the week containing Easter Sunday, and then through the week following Easter Sunday. Our visual analysis has made an interesting discovery. This is, of course, a simple example, and an egg manager would already know this and not need to go through this analysis. For us, starting with simple examples

Figure 4.11: Line Plot of Egg Case Sales over Weeks

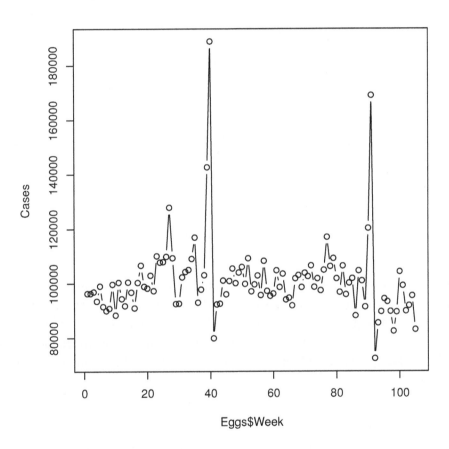

Figure 4.12: Boxplot Dialog

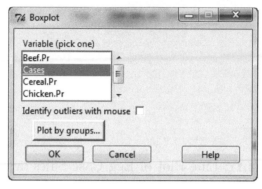

Figure 4.13: Boxplot Group Variable Selection

allows us to practice using these methods so that we can gain the necessary skills to apply them to more complex situations.

Boxplots

R provides a number of useful tools for visualizing the relationship between a continuous variable such as "Cases," and a categorical variable such as "Easter." One of the best tools for doing this is known as a box plot. You can create a box plot by using the pull-down menu option **Explore and Test** → **Visualize**→**Boxplot...**, which will cause the dialog box shown in Figure 4.12 to appear.

7.

In the **Boxplot** dialog box **select Cases as the "Variable (pick one),"** and then click on the **Plot by groups...** button, which will cause the dialog box shown in Figure 4.13 to appear, with any categorical variables that are in the data set that could be used as grouping variables.

8.

Select Easter as the "Groups variable (pick one)" in the **Group** dialog box, and then press **OK**. This will bring you back to the **Boxplot** dialog box,

Figure 4.14: The Revised Boxplot Dialog

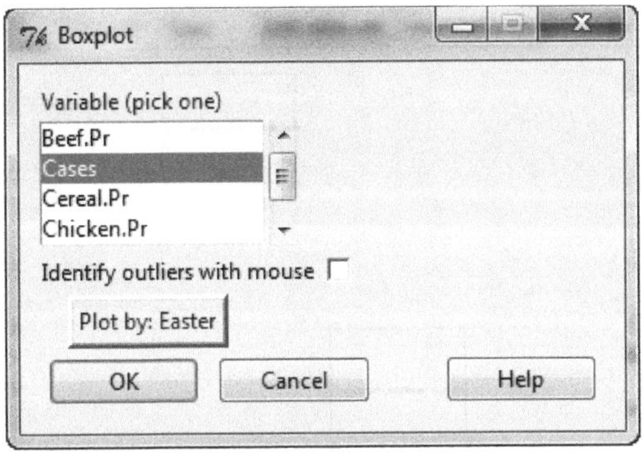

but with the **Plot by groups...** button now indicating that Easter is the grouping variable (Figure 4.14). **Press OK in the Boxplot dialog box**, and the box plot shown in Figure 4.15 will be created.

9.

By selecting a categorical grouping variable, the resulting multiple boxplots provide a way of comparing a *continuous* variable (Cases) with a *categorical* variable (Easter). The boxplot on the left is more detailed than the other three, and the most typical type of boxplot. It corresponds to weeks not in the Easter season (i.e., when Easter takes on the level "Non Easter"). The boxplot shows the distribution of a single variable, and is somewhat like a simplified histogram (see the histogram tutorial in Chapter 3). It also highlights the outliers in the distribution of values of the variable. The line towards the middle of the "box" corresponds to the median egg case sales at a given level of Easter. For the Non Easter sales, the median is just below 100,000 (half of the weeks have sales greater than 100,000, half have less). The main box corresponds to the range of the 25th to 75th percentile of Cases, or the middle two quartiles, which corresponds to case sales of about 94,000 (one quarter of the weeks have sales less than 94,000) and 103,000 (one quarter of the weeks have sales greater than 103,000). The whisker, or stem, that extends beyond the box goes out to the largest and smallest data values, unless those values are more than 1.5 times the width of the box. The height of the box corresponds to what is known as the "interquartile range" (greater than the lower quarter, less than the upper quarter). For Non Easter, the lowest sales are at about 82,000 cases. Any points that are more than 1.5 times the interquartile range away are represented by individual circles. The highest sales for Non Easter weeks is about 128,000 cases, which is represented by the small circle in the boxplot. This is the only week in which sales are more than 1.5 times the box width (interquartile range). The box plots for the Pre Easter, Easter, and Post

Figure 4.15: Boxplot of Egg Case Sales Grouped by Easter Weeks

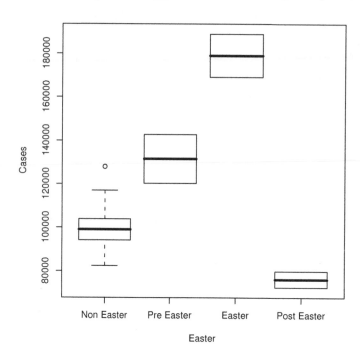

Easter values of the categorical `Easter` variable are unusual in this instance in that they don't have whiskers. The reason for this is that there are only two data points for each. With only two years of data, we only observe two Easter seasons, and hence a 25th to 75th percentile range is meaningless. The median lines, however, are meaningful, and clearly demonstrate the enormous effect that the Easter season has on egg sales. Sales during these three weeks lie far outside the usual range of sales for non-Easter weeks.

10.

Exercise: At this point you should check the relationship between the other variables included in the data set and case sales of eggs. Starting with the categorical variables, create a box plot of `Cases` for different levels of the variable `First.Week` (a variable that indicates whether an observation corresponds to the first week in a month). What are the median and quartile values (approximately) shown in the boxplots? Does `First.Week` appear to have an effect? What is that effect? Can you think of a plausible reason why this effect might exist?

Scatterplot Matrix

A tool that allows a quick exploration of the relationship among several continuous variables is the Scatterplot Matrix. **Select Explore and Test → Vi-**

Figure 4.16: Scatterplot Matrix Dialog

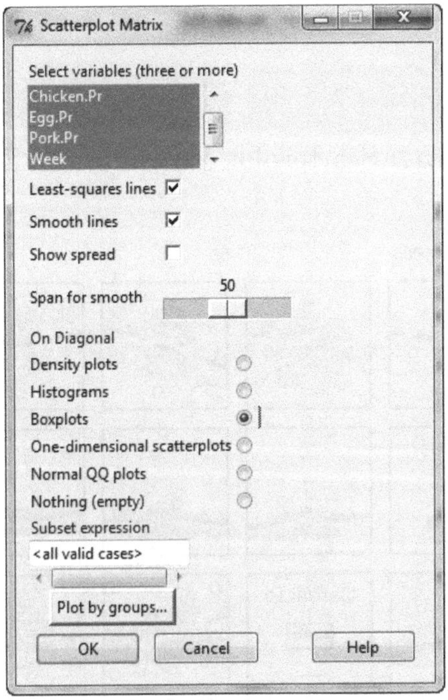

Scatterplot Matrix

Select variables (three or more)
Chicken.Pr
Egg.Pr
Pork.Pr
Week

Least-squares lines ☑

Smooth lines ☑

Show spread ☐

Span for smooth 50

On Diagonal
Density plots ○
Histograms ○
Boxplots ◉
One-dimensional scatterplots ○
Normal QQ plots ○
Nothing (empty) ○
Subset expression
<all valid cases>

Plot by groups...

OK   Cancel   Help

sualize → Scatterplot Matrix... to bring up the dialog box in Figure 4.16. **Select all seven of the continuous variables, and the boxplot option. Click OK.** The result in the graphics window will be a matrix of scatterplots for all combinations of the seven variables (Figure 4.17). This is close to the maximum number of variables that can be easily interpreted in the scatterplot matrix. The diagonal shows the boxplots for each variable. Horizontally along a row from the boxplot shows the same variable as the y-variable in the scatterplots; for example, the second row has Cases as the y-variable, with the other six variables as x-variables. The second row and fifth column, for example, shows the Egg prices vs. Cases scatterplot we examined earlier.

Exercise: Identify any other variables related to Egg Case sales, and speculate as to why the relation(s) might exist.

11.

Linear Regression

Having looked at the data visualization tools for continuous dependent variables, we now move on to an examination of R's linear regression tools. To estimate a multiple (more than one predictor) linear regression model use the

Figure 4.17: Scatterplot Matrix of the Eggs Data

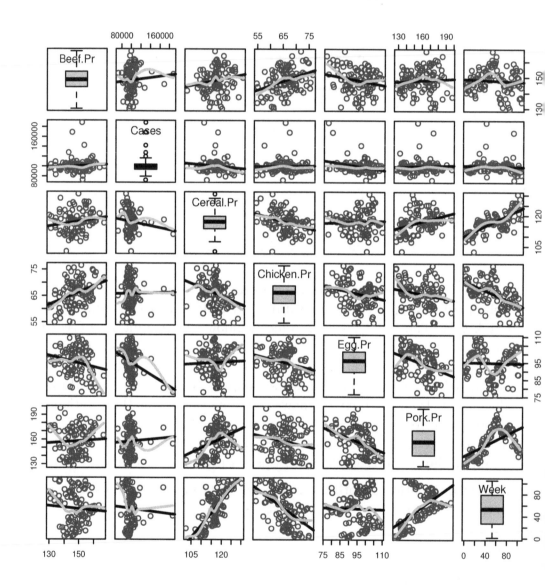

Figure 4.18: The Linear Model Dialog

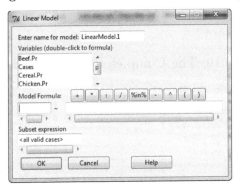

pull-down menu option **Models → Statistical models → Linear model...** to bring up the Linear Model dialog box, shown in Figure 4.18.

12.

By default, R Commander names a linear regression model `LinearModel.n`, where n is the sequential number of models estimated during that R Commander session (thus the next model we estimate will be given the default name `LinearModel.2`). This is a very useful feature of R Commander, since it prevents us from writing over existing models by mistake. However, we are likely to prefer a more descriptive name. Since in the first model we will not be transforming any of the variables, the model we estimate will be truly "linear." Consequently, it makes sense to name this model LinearEggs since it indicates both the structure and the data set used in constructing the model, and **enter it in the "Enter name of model:" field.** Next **double-click on the variable `Cases`** in the "Variables (double-click to formula)" scroll box. By default, the first variable you double-click from this scroll box is selected as the dependent variable of the model. **Next double-click, in turn, the variables `Beef.Pr`, `Cereal.Pr`, `Chicken.Pr`, `Easter`, `Egg.Pr`, `First.Week`, `Month`, `Pork.Pr`. Do not select `Week`.** You have to select each variable individually, and cannot select multiple variables at once. Each variable will automatically be separated by a plus sign (+). R Commander also recognizes categorical (factor) variables like Easter and month, and automatically creates the necessary indicator (dummy) variables. When you have completed all of this, your dialog box should look like the one shown in Figure 4.19. When it does, **press OK**, and the results of the estimated model will appear in R Commander's output window (Figure 4.20). This window shows the usual regression ouput information, including coefficient estimates, t- and p-values, and the goodness of fit statistics R^2 and adjusted R^2.

Figure 4.19: The Completed Linear Model Dialog

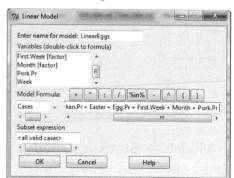

Figure 4.20: Linear Regression Results for LinearEggs

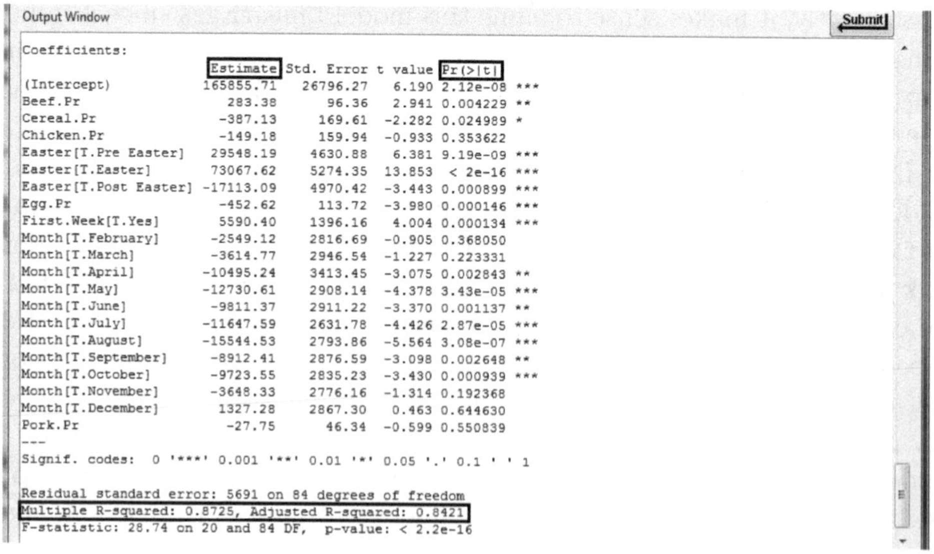

```
Output Window                                                               Submit

Coefficients:
                      Estimate  Std. Error  t value  Pr(>|t|)
(Intercept)          165855.71    26796.27    6.190  2.12e-08 ***
Beef.Pr                 283.38       96.36    2.941  0.004229 **
Cereal.Pr              -387.13      169.61   -2.282  0.024989 *
Chicken.Pr             -149.18      159.94   -0.933  0.353622
Easter[T.Pre Easter]  29548.19     4630.88    6.381  9.19e-09 ***
Easter[T.Easter]      73067.62     5274.35   13.853  < 2e-16  ***
Easter[T.Post Easter] -17113.09    4970.42   -3.443  0.000899 ***
Egg.Pr                 -452.62      113.72   -3.980  0.000146 ***
First.Week[T.Yes]      5590.40     1396.16    4.004  0.000134 ***
Month[T.February]      -2549.12     2816.69   -0.905  0.368050
Month[T.March]         -3614.77     2946.54   -1.227  0.223331
Month[T.April]        -10495.24     3413.45   -3.075  0.002843 **
Month[T.May]          -12730.61     2908.14   -4.378  3.43e-05 ***
Month[T.June]          -9811.37     2911.22   -3.370  0.001137 **
Month[T.July]         -11647.59     2631.78   -4.426  2.87e-05 ***
Month[T.August]       -15544.53     2793.86   -5.564  3.08e-07 ***
Month[T.September]     -8912.41     2876.59   -3.098  0.002648 **
Month[T.October]       -9723.55     2835.23   -3.430  0.000939 ***
Month[T.November]      -3648.33     2776.16   -1.314  0.192368
Month[T.December]       1327.28     2867.30    0.463  0.644630
Pork.Pr                  -27.75       46.34   -0.599  0.550839
---
Signif. codes:  0 '***' 0.001 '**' 0.01 '*' 0.05 '.' 0.1 ' ' 1

Residual standard error: 5691 on 84 degrees of freedom
Multiple R-squared: 0.8725, Adjusted R-squared: 0.8421
F-statistic: 28.74 on 20 and 84 DF,  p-value: < 2.2e-16
```

13.

Interpretation

In Figure 4.20, R^2 indicates that the model fits very well (it explains over 87% of the variance in retail egg sales). In addition, the results also indicate that many of the included variables are highly statistically significant based on their p-values (the null hypothesis of these tests is that the coefficients equal zero).

Exercise: First use the statistical results: Which of the variables seem to be most important? In particular, what do the coefficients on the monthly dummy variables suggest about the seasonal pattern of egg sales? What is the impact of an increase in the price of eggs on retail egg sales? An increase in the price of beef? An increase in the price of breakfast cereal? Now use your judgment: Do all of these results make logical sense? That is, are the signs (+ or −) and magnitudes of the coefficients consistent with your expectations?

14.

Interpreting Categorical Predictors

One issue with variables that are factors (categorical) is that each *level* of a factor has its own estimated coefficient that measures the impact of that level relative to the (omitted) "base case" level of that factor. For instance, for Month the omitted level is "January," and the parameters for the other 11 months indicate the *difference* (and whether it is statistically significant) between that month and January. For example, 15,544 fewer cases, on average, are sold in August than in January. As a result, different choices of the omitted level will alter the pattern of estimated coefficients and significance levels, although the choice of the omitted level has no influence on overall model fit or, if we use the model for prediction, the resulting predicted values of Cases sold. If we look across the levels of Month in the estimated model we find that some of the coefficients are negative, while others are positive. Similarly, some of the coefficients (and hence the difference in the effect from January) are statistically different from zero, while others are not. Finally, based on these results, we do not know, taken as a whole, whether Month is statistically significant. R Commander provides a very nice tool to assess whether, taken as a whole, a factor variable is statistically significant. To use this tool use the pull-down menu option **Assess →Hypothesis tests →ANOVA table**, and **press OK**, which will cause the results shown in Figure 4.21 to be printed to the R Commander output window. Examining the results in Figure 4.21 you can see that, taken as a whole, all three factor variables (Easter, First.Week, and Month) in this model are all highly statistically significant.

Figure 4.21: ANOVA Table Hypothesis Test

```
Output Window

> Anova(LinearEggs, type="II")
Anova Table (Type II tests)

Response: Cases
                 Sum Sq Df F value    Pr(>F)
Beef.Pr       280103188  1  8.6478 0.0042295 **
Cereal.Pr     168742668  1  5.2097 0.0249886 *
Chicken.Pr     28180179  1  0.8700 0.3536222
Easter       8773987097  3 90.2948 < 2.2e-16 ***
Egg.Pr        513078180  1 15.8406 0.0001457 ***
First.Week    519311341  1 16.0330 0.0001338 ***
Month        2107016293 11  5.9137 4.886e-07 ***
Pork.Pr        11618467  1  0.3587 0.5508388
Residuals    2720772383 84
---
Signif. codes:  0 '***' 0.001 '**' 0.01 '*' 0.05 '.' 0.1 ' ' 1
```

15.

Nonlinear Model

Having estimated the linear model specification, we will next estimate the multiplicative power function specification. This specification assumes that, instead of **Cases** being a function of the sum of the predictor variables multiplied by a coefficient (that is estimated), **Cases** is the product of the predictor variables, each raised to the power of an estimated coefficient. This is easy to do, because if we take the natural logarithm of each side of the multiplicative power function model, it becomes a linear model of logs of the variables, and we can use standard linear regression to estimate the coefficients. To estimate the power function specification we must, therefore, apply a natural logarithm transformation to the dependent variable (**Cases**) and all the continuous (ratio or interval scaled) predictor variables (i.e., the five price variables). However, we do not apply the natural logarithm transformation to the three factor variables. In order to create new variables based on the appropriate variable transformations use the pull-down menu option **Data → Manipulate variables → Compute a new variable...**, which will bring up the dialog box shown in Figure 4.22.

16.

In the "New variable name" field enter the name you wish to give to the log transformed **Cases** variable, which can be any name you like. In Figure 4.22 the name **Log.Cases** has been typed in. The next box specifies the computation, which is to take the logarithm of each value of the Cases variable, and has to be

Figure 4.22: Compute New Variable Dialog

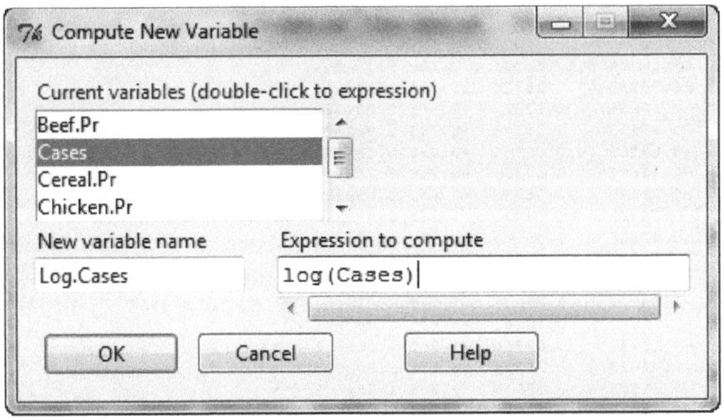

entered exactly as shown. Enter the formula `log(Cases)` into the "Expression to compute" field, and then press **OK** to create the new variable `Log.Cases`.

17.

Repeat steps 15 and 16 using the five price variables (changing variable names) to create the five natural log transformed price variables. Check that all of the variables have been created by **View Data**. One you have done this, repeat the regression analysis, only this time using the new log transformed variables rather than the original case and price variables (reviewing step 12 may help here). Specifically, this model should use the log transformed **Cases** and price variables along with the original three factor variables. This is equivalent to estimating the power function model of the original variables. One thing to be aware of is that when you start, the **Linear Model** dialog box will have your previous model initially entered. This is often useful since a common thing to do during an analysis is to add or remove variables from the previous model estimated. Although, in this case we are altering so many variables that it makes more sense to delete everything in the two model fields and start over. Once you have estimated the model, your results should look like those contained in Figure 4.23.

18.

Compare the results of the two estimated models. **Which fits the data better?**[10] **Exercise:** How do the estimated coefficients differ across these two models, particularly their signs and significance? Although the magnitudes are

---

[10]A technical point is that the adjusted R^2 values for the logged and unlogged dependent variables are not strictly comparable; however, the differences are usually minor. In subsequent chapters we introduce a different and better method of comparing models using holdout samples, which avoids the problem.

Figure 4.23: Linear Regession Results for the Power Function Model

```
Output Window                                                                    Submit

Coefficients:
                     Estimate Std. Error t value Pr(>|t|)
(Intercept)          13.90880    1.18826  11.705  < 2e-16  ***
Log.Beef.Pr           0.39135    0.13520   2.895 0.004836  **
Log.Cereal.Pr        -0.40900    0.18764  -2.180 0.032076  *
Log.Chicken.Pr       -0.06125    0.09886  -0.620 0.537223
Log.Egg.Pr           -0.42666    0.10114  -4.218 6.18e-05  ***
Log.Pork.Pr          -0.03139    0.06936  -0.452 0.652098
Easter[T.Pre Easter]  0.26223    0.04401   5.958 5.77e-08  ***
Easter[T.Easter]      0.52815    0.05060  10.438  < 2e-16  ***
Easter[T.Post Easter]-0.19409    0.04712  -4.119 8.86e-05  ***
Month[T.February]    -0.02205    0.02681  -0.823 0.413119
Month[T.March]       -0.03116    0.02806  -1.110 0.270009
Month[T.April]       -0.11087    0.03241  -3.421 0.000966  ***
Month[T.May]         -0.12647    0.02756  -4.588 1.55e-05  ***
Month[T.June]        -0.09423    0.02759  -3.416 0.000983  ***
Month[T.July]        -0.11508    0.02499  -4.604 1.46e-05  ***
Month[T.August]      -0.15438    0.02661  -5.802 1.13e-07  ***
Month[T.September]   -0.08326    0.02734  -3.046 0.003102  **
Month[T.October]     -0.08994    0.02704  -3.326 0.001308  **
Month[T.November]    -0.03067    0.02636  -1.164 0.247888
Month[T.December]     0.01695    0.02723   0.623 0.535248
First.Week[T.Yes]     0.05584    0.01327   4.207 6.46e-05  ***
---
Signif. codes:  0 '***' 0.001 '**' 0.01 '*' 0.05 '.' 0.1 ' ' 1

Residual standard error: 0.05409 on 84 degrees of freedom
Multiple R-squared: 0.8396, Adjusted R-squared: 0.8014
F-statistic: 21.99 on 20 and 84 DF,  p-value: < 2.2e-16
```

very different between the two models, are the larger magnitudes in one model
the same as the larger magnitudes in the other? Which variables appear to
have effects that are "robust" to the selection of a particular functional form
(that is, remain fairly similar in sign and relative magnitude and significance
in both models)?

19.

A Good Model

Exercise: Using these results, do some exploratory data analysis, **with the
objective of estimating a model which minimizes the number of
predictor variables used in the model, but retains a fairly high R^2**.
For continuous predictors, try omitting non-significant predictors. The R^2 will
typically drop with each variable removed, but will drop more with some
variables than with others. An additional statistic reported that helps with
the decision on which variables to retain and which to drop is Adjusted R^2.
This number is calculated by subtracting an amount from R^2 which depends
on the number of variables used in the regression. The more variables used,
the more the R^2 is reduced by the adjustment. Adding variables will almost
always increase the R^2, simply by the small bit of fit improvement that a purely
random relation will generate; but may have little effect on the Adjusted R^2.
If the increase in R^2 is very small, the Adjusted R^2 will decrease with the
addition of variables. In essence, the adjustment attempts to compensate for

purely random fit improvements, but, unfortunately, does not completely do so. **Still, finding a model with the largest Adjusted R^2 is a good approach.**

For factor variables, try to relabel factor levels of some of them using the tools introduced in the tutorials of the last chapter in order to decrease the number of levels of the variable (e.g., go from a level for each month in a year to a smaller number of "season" levels). Put your resulting best model in a word document, and briefly describe how you arrived at it.

Chapter 5

Logistic Regression

In the last chapter we provided an introduction to linear regression, the first predictive modeling tool created, and one that is still extremely powerful. Unfortunately, the method does have its limitations. One important one is that categorical target variables (variables with a nominal or ordinal scale) violate the assumptions underlying linear regression. However, in a pinch, a reasonable predictive model for a binary target variable (typically of the "Yes/No" variety) can be created using linear regression (assuming the values of "Yes" and "No" have been converted to values of 1 and 0). The same is true for an ordinal target variable, assuming the categories of the variable have been given numerical values that reflect their ordinal positions. Even though it is possible to create reasonable predictive models for binary and ordinal target variables, more appropriate methods exist.

One problem with using linear regression for a binary target variable is that what we are interested in predicting is the *probability* that a customer will respond in a favorable way (e.g., that his or her response will be "yes" to our offer). The predicted probability of a favorable response should fall between zero and one. However, using linear regression, we will often predict probabilities that fall outside the zero to one range.

Logistic regression is the preferred method for developing a regression-like predictive model when the target variable is binary. The method falls in the same broad class of methods (known as generalized linear models) as linear regression, and provides a similar set of outputs, so if you are comfortable reading a linear regression output, a logistic regression output is also easy to read. In this chapter we again begin by providing a graphical visualization of a one predictor problem along with the logistic regression approach, to build some intuition for the method, and then move on to explaining some of the technical details behind the method. The final section is devoted to a tutorial that covers visualization methods for examining the relationship between a binary target variable and several potential predictor variables, estimating logistic regression models, and assessing the model.

5.1 A Graphical Illustration of the Problem

To motivate the illustration, consider the case of a charity that has developed a frequent giving program in which donors make a small, pre-determined donation that is automatically debited against a donor's credit card each month. Since the amounts are small, this has a limited effect on the donor's monthly cash flow, allowing that donor to comfortably give more in total than is possible with a single annual donation. A trial program done by the charity reveals that people who do enroll in the frequent giving program do in fact give substantially more than they did before entering the program. The question now is whom, in the charity's database, to target with an offer to join this monthly giving program in order to increase the percentage of the donor base that are enrolled in the program?

A preliminary analysis done by the charity reveals that one important factor in determining whether a donor joins the frequent giving program is that donor's average annual donation amount prior to receiving an offer to join the frequent giving program. Figures 5.1 and 5.2 provide a scatterplot of the relationship between whether a donor joined the program (given a value of 1)

Figure 5.1: Joining the Frequent Donor Program and Average Annual Donation Amount, Database 1

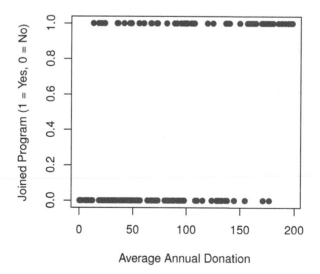

Figure 5.2: Joining the Frequent Donor Program and Average Annual Donation Amount, Database 2

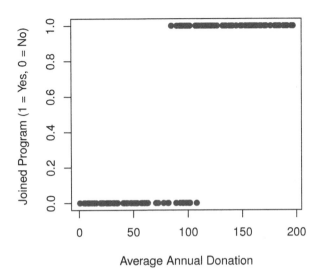

or not (given a value of 0) and each donor's average donation amount for two different hypothetical databases.

An examination of the two figures indicates that the relationship between joining the program and past average donation amounts is greater in the second example database than it is in the first. However, in both cases it is apparent that as a donor's past average donation amount increases, the more likely that donor is to join the monthly giving program.

Ideally, we want to have an equation that maps the level of the past average annual donation amount to the probability of joining the frequent giving program for donors in the charity's database in order to determine which donors should be targeted with an offer to join the program. Figures 5.3 and 5.4 illustrate what this relationship would look like graphically for the two different hypothetical databases. In the figures the vertical (y) axis of the plots give the probability of joining the frequent giving program for a given past average annual donation amount. Consistent with what one would expect, the relationship between the average annual donation amount and the probability of joining the frequent giving program is sharper in the second example than it is in the first.

Logistic regression provides a means of estimating these probability functions. While our example has been based on examining a single predictor relation-

Figure 5.3: Probability Plot for Database 1

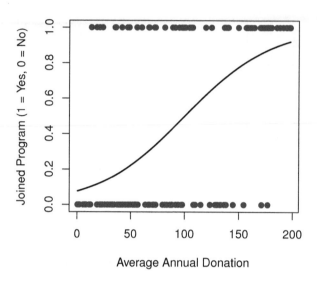

Figure 5.4: Probability Plot for Database 2

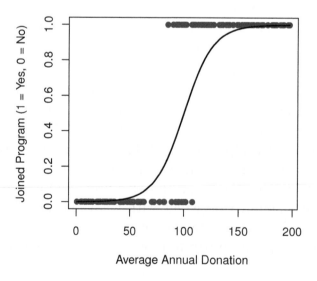

ship, logistic regression (like linear regression) can use more than one predictor variable, and some or all of those predicator variables can be indicator variables. As indicated earlier, there are a number of similarities between linear and logistic regression, the topic to which we now turn.

5.2   The Generalized Linear Model

The generalized linear model was proposed in the early 1970s by Nelder and Wedderburn (1972) in an effort to provide a means of using linear regression-like models to address problems that were not directly amenable to the application of linear regression.[1] The class of models they proposed included linear and logistic regression models as special cases, as well as a number of other related models. The fundamental equation of the generalized linear model is

$$g(\mathrm{E}(y)) = \alpha + \beta x_1 + \gamma x_2,$$

where $\mathrm{E}(y)$ is the expectation of the target variable y, $\alpha + \beta x_1 + \gamma x_2$ is the linear predictor (the coefficients of which, α, β, and γ, are to be estimated), and $g()$ is the link function that maps the expectation of y to the linear predictor. While we present only the use of two predictor variables here, you should be aware that the method generalizes to an arbitrary number of predictors.

For our purposes, it is easier to think of things in terms of the inverse link function (which we label $f()$), rather than the link function itself. Consequently, we are interested in the equation

$$\mathrm{E}(y) = f(\alpha + \beta x_1 + \gamma x_2).$$

In the case of linear regression, the inverse link function is simply the identity function, so this equation becomes

$$\mathrm{E}(y) = \alpha + \beta x_1 + \gamma x_2.$$

Things become more complex for logistic regression since the expectation of the target variable y is actually a probability (which we label $\mathrm{Pr}(y = 1)$, but it could be $y = $ "Yes," or any other value we labeled as the desired outcome for the target variable), so the inverse link function needs to map the linear

---

[1] McCullagh and Nelder (1989) is an excellent, and updated source, on the generalized linear model.

predictor (which can take on any real value) to a value that falls in the zero to one interval of a probability. Moreover, that transformation must be an increasing monotone transformation of the linear predictor (which means as the value of the linear predictor increases, so does the probability). While these two conditions greatly limit the number of potential link functions, there is still a number of functions that meet these requirements. The most commonly used inverse link function is

$$E(y) = \Pr(y = 1) = \frac{\exp(\alpha + \beta x_1 + \delta x_2)}{\exp(\alpha + \beta x_1 + \delta x_2) + 1}.$$

This inverse link function corresponds to the cumulative density function of the logistic distribution, which gives the method its name. Frequently, a logistic regression model using this inverse link function is called a *logit* model. Examine this function more carefully (something that might seem a bit daunting at first, but may be less difficult than it seems). In particular, notice that as the linear predictor takes on extremely large *negative* values that the term $\exp(\alpha + \beta x_1 + \delta x_2)$ approaches zero, so the ratio (the probability) also approaches zero. Conversely, when the linear predictor takes on extremely large positive values, the term $\exp(\alpha + \beta x_1 + \delta x_2)$ approaches positive infinity, so the ratio approaches one. The probability curves shown in Figures 5.3 and 5.4 are based on this inverse link function.

The other commonly used inverse link function is the cumulative density function of the standard Normal distribution (typically written as $\Phi(\alpha + \beta x_1 + \delta x_2)$), and a logistic regression model that uses this inverse link function is commonly referred to as a *probit* model. The drawback to the probit model is that the cumulative density function of the standard Normal distribution is not "closed form," which means (unlike for the cumulative logistic density function) values for it must be found via numerical integration (making this model computationally more demanding than the logit model). As illustrated in Figures 5.5 and 5.6 for the two examples of the frequent giver program used earlier, the value of the probit inverse link function for a particular value of the linear predictor (which is set to an average annual donation amount of $125 in the figures) is equal to the area under the standard Normal density curve from minus infinity to $125. As can be seen in the figures, the density is much more concentrated for the second example database, so the cumulative density (the probability of joining the monthly giving program) when the past annual donation amount is $125 is greater in this database than it is in the first database.

Figure 5.5: The Probit Inverse Link Function for Database 1

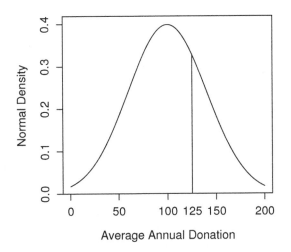

Figure 5.6: The Probit Inverse Link Function for Database 2

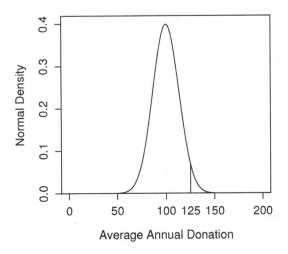

5.3 Logistic Regression Details

The logistic regression algorithm works by finding the set of coefficient values (α, β, and γ for our example linear predictor) that maximizes the function

$$\prod_i [f(\alpha + \beta x_{1i} + \gamma x_{2i})]^{y_i} [1 - f(\alpha + \beta x_{1i} + \gamma x_{2i})]^{(1-y_i)}.$$

This function is known as a likelihood function (since we are selecting coefficient values that maximize the likelihood of explaining the observed data). We admit that you might find this function to be a bit intimidating. However, it is actually less complicated than it appears. The index i is a counter for all customers/donors in the database, while the term $\prod_i$ is the product operator (i.e., it just indicates that the values for all the customers/donors in the database are multiplied together). The other major change is that now the target variable and the predictor variables are indexed by record (i.e., the value of y_i, x_{1i}, *and* x_{2i} are the values of these variables for record i of the database).

As you may have already guessed, $f(\alpha + \beta x_{1i} + \gamma x_{2i})$ is the probability that customer/donor i will respond favorably, while y_i indicates whether customer/donor i did in fact respond favorably. If the model is exactly correct for donor i (who responded favorably), then both y_i and $f(\alpha + \beta x_{1i} + \gamma x_{2i})$ will equal one, as will the value of the likelihood for that consumer/donor. If, on the other hand, the model is exactly wrong, then y_i will equal one, while $f(\alpha + \beta x_{1i} + \gamma x_{2i})$ will equal zero, and the value of the likelihood for that customer/donor will equal zero (which would cause havoc given the multiplicative structure of the likelihood function since one exactly wrong value would make the entire value of the likelihood equal zero, but this never actually occurs). Overall, the likelihood function indicates a good fit as its value approaches one, and a poor fit of the data as its value approaches zero.

Numerical optimization methods cannot maximize the value of a product, but they can minimize the value of a sum. As a result, the natural logarithm of the likelihood function is taken (converting it into a sum rather than a product) which is a monotone transformation of the likelihood (so it has the same maximum point as the likelihood itself), and is multiplied by a negative number (which reaches its minimum for the same coefficient values which maximize the original likelihood). In order to ease the computation of hypothesis test statistics for the estimated logistic regression model (but without influencing the coefficient estimates), the natural logarithm of the likelihood function is multiplied by -2 (rather than -1 as one might expect), resulting in a quantity that is called the *deviance* (which we admit is an odd name). Thus, logistic regression ultimately attempts to find coefficient values that minimize the deviance of a model given the data, but these same coefficient values correspond to those that maximize the original likelihood function. A model that fits the data perfectly (something that never occurs) would have a deviance of zero, while the deviance for a less-than-perfect fit is greater than zero. There are a number of methods that can be used to minimize the deviance of the model given the data, and R uses a method known as Fisher Scoring (McCullagh and Nelder, 1989).

While the important statistics generated in the estimation of a logistic regression model are not the same as they are for a linear regression model, they are analogous. Specifically, a z-test (based on the standard Normal distribution) is used to test whether the individual coefficients are different from zero rather than a t-test, but the p-values of these tests are interpreted in exactly the same way as they are in linear regression.

A true R^2 cannot be calculated, but a McFadden R^2 can be calculated (McFadden, 1974). As with R^2, the McFadden R^2 is bounded between zero and one, and the closer this statistic is to one, the better the model fits. However, as with R^2, the value of the McFadden R^2 never decreases (and typically increases) as additional variables are added to the model, leading to potential overfitting problems. The formula used to calculate the McFadden R^2 value is

$$1 - \frac{deviance}{null\ deviance},$$

where the *null deviance* is the deviance of a logistic regression model that contains only a constant. If the model with additional variables does no better than a constant-only model, then the deviance and null deviance are equal to each other, and the McFadden R^2 equals zero. On the other hand, if the model with additional variables fits perfectly, then the deviance of that model equals zero, and the McFadden R^2 equals one. To give some sense of what different McFadden R^2 values mean, the probability function shown in Figure 5.3 has a McFadden R^2 of 0.25, while the probability function in Figure 5.6 has a McFadden R^2 of 0.6.

In linear regression the use of adjusted R^2 is intended to mitigate potential overfitting problems associated with R^2. The analogous statistic for logistic regression is known as the Akaike Information Criterion (or AIC; Amemiya, 1985), which is given by

$$deviance + 2c,$$

where c is the number of estimated coefficients in the model. Based on this formula, it is clear that the AIC is the value of the deviance for a model that has been penalized for the number of model coefficients. Given this structure, we are often interested in finding the model that has the *minimum* AIC value. At this point it should be noted that the AIC can also be used for linear regression models as well since the criterion used in linear regression (minimizing the sum of the squared model errors) corresponds to the minimization of the deviance of a model if we assume that the error terms are normally distributed. Since the deviance of a linear regression model can be calculated, so can its AIC. However, by historical convention, adjusted R^2 is the statistic usually reported.

5.4 Logistic Regression Tutorial

The purpose of this tutorial is to make you familiar with some of the logistic regression tools in R. In addition a number of the visualization tools that were examined in the linear regression tutorial (Chapter 4) will also be used in this tutorial. It turns out that visualizing the relationship between potential predictor variables and the target variable requires a bit more work when the target variable is a categorical (a *factor*, in R) rather than a continuous variable. Finally, you will see an example of a process that can be used to quickly develop a reasonably good predictive logistic regression model. It is likely that a better model could be found, but the improvement would be small, and the effort large.

5.4.1 Highly Targeted Database Marketing

The data set used to illustrate the methods presented in this lab involves a variant on an "up-selling" application. Specifically, the data come from a project conducted for a Canadian Charitable Society, British Columbia and Yukon Region (or CCS). The objective of this project was to develop a model to determine which of the current CCS donors should be selected for a targeted mailing to encourage them to join the "monthly giving program." Under the monthly giving program a donor to the CCS elects a monthly amount to give and provides a credit card number so that automatic monthly donations are made. The CCS had offered this program as one of several payment options under its general annual appeal campaign over the prior three years. Donations from individuals who had moved to this program from the annual campaign increased their annual giving by a factor of three on average (e.g., moving from giving the CCS $20 a year to $60 a year), and had a very low attrition rate (about a 3 percent loss per year compared to the 33 percent loss for the annual campaign). However, only about 1 percent of donors elected to join the monthly giving program when it was offered as one of several payment options and was given no special prominence in the CCS's appeal. Based on this experience, the CCS was interested in developing a special campaign to increase the number of donors enrolled in the monthly giving program. In the absence of any modeling, all that the CCS could do would be a costly mass mailing with an attendant low response rate. A highly targeted campaign would be preferable, requiring identifying those current donors most likely to upgrade to the monthly giving program in response to CCS's direct mail.

To this end, a database was constructed that included donors' behaviors prior to being offered the monthly giving program, and a variable indicating whether or not they had joined the program after the offer. A simple analysis would be to compare the joiners and non-joiners on the variables available and look for differences. A much better approach would be to build a model of the probability of joining the program as function of available variables, and then to use this model to predict the probability of other donors accepting the offer to join the monthly giving program.

In the following subsections, we will first introduce two important concepts in database marketing, over-sampling and controlling over-fitting.

5.4.2 Oversampling

Because the percentage of individuals who belonged to the monthly giving program is a very small percentage of the total donor database, the large majority of non-members will overwhelm any predictive model estimated on the total database. Determining what influences an individual to become a member of the program is therefore difficult. Instead, following common data mining practice, samples were drawn (without replacement) to create a new database that consists of a 50–50 percent split of monthly givers and non-monthly givers. This is known as *over-sampling* the target population of monthly program members. A model estimated on this more balanced database has a better chance of recognizing differences between the two types of donors.[2] When using the final model to calculate return on investment, the important trick to remember is that we have over-sampled the members of the program to create the model. We will need to correct from the one-in-two proportion in the oversampled database to the 1-in-100 proportion in the original CCS database. That is a simple ratio correction, and will be discussed later.

---

[2]The needed level of over-sampling varies with the predictive modeling tool used. If only logistic regression is used, over sampling to 20–80 split seems to work fairly well. However, other predictive modeling tools (tree models in particular) work best with a 50–50 split sample.

5.4.3   Overfitting and Model Validation

In the previous discussion, we really should say "samples" (as opposed to "sample"). The best way to control overfitting using any of the models we will encounter from here on is to split the database into two (sometimes more) samples. Each contains the same variables and oversampling rate, but included completely different individuals. An *estimation sample* is used to calibrate several potential models. The *validation sample* is not used in the process. It is held back, and used to test, or validate, the models, and to select the best among them. This is done by using the predictor variables from the validation data (not used to create the model) to calculate *predicted* target values. Since we also have the *observed* target values, we can compare, and determine how well the model predicts on data that were *not* used in creating the model. We then select the model (estimated using the estimation sample) that best predicts the target value in the validation sample. Some modeling procedures take this one step further, creating a third sample, a *holdout sample*. In the remainder of this book we will only use two samples.[3]

This procedure is used to avoid the possibility of developing a model that happens to fit the estimation sample data well but is fitting variations that are not repeatable from sample to sample, that is, fitting unrepeatable noise in the data. When applied to the new validation sample with different "noise" it will fit worse. We have already met this problem, and one way of helping to control for it linear regression. The adjusted R^2 controls for merely capturing noise when adding variables by reducing the R^2 by an adjustment that depends on the number of variables used. The adjusted R^2, the AIC, and validation and holdout methodologies, guard against *overfitting* the estimation sample.

In this tutorial, and those in the next three chapters, we are interested in calibrating and selecting models, so we will only be working with the estimation and validation samples, which we will create using tools in R Commander. The data set we will be using in this tutorial (and for the next several) is entitled CCS, and can be found in the BCA data library.

1.

Use the pull-down menu command **Data → Get From →R package → Read data from an attached package. . .** to read the **CCS** data set from the BCA package into R.

---

[3]Jargon Alert: The estimation sample is often called the *training* sample, and the validation sample is sometimes called a *test* sample. As well, when only two samples are used, the terms "holdout" and "validation" can both be applied to the second sample. Always take care to clarify exactly what is meant when you see these terms.

Figure 5.7: The Create Samples Dialog

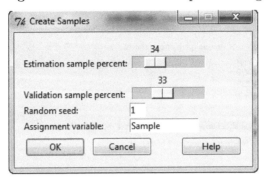

2.

Before proceeding further, we will create two different samples of individuals within the data, an estimation and validation sample. In this tutorial we will estimate several logistic regression models using the estimation sample, and in the tutorial of the next chapter we will use the validation sample to assess whether our models can accurately predict new data. To create the samples use the pull-down menu command **Data → Organize → Create samples in active data set...**, which brings up the dialog box shown in Figure 5.7. As indicated above, for this tutorial, and the next several, we will only be using an estimation and validation sample, but not a holdout sample. With this dialog, the size of the holdout sample is not specified, but is determined by the percentage of the total data set not included in either the estimation or validation sample. In Figure 5.7, 77% of the sample is for estimation and validation, leaving 33% for holdout. Consequently, if the percentage of the total data set allocated to the estimation and validation samples equals 100 percent, no holdout sample is created. At this point **use the slider bars to set both the estimation and validation samples to 50 percent of the data set**. The two samples are drawn at random (without replacement). Which specific records are allocated to each data set is determined the "Random seed:" field, which controls how R generates random numbers. The use of a random seed allows the same samples to be created by different users on different occasions. **Use the View data set button** to see that a new variable named "Sample" has been added to the data set, with values "Estimation" and "Validation." This variable will be used to specify which individuals we use for the following procedures.

3.

As in the linear regression tutorial, we will start with an examination of data visualization tools. The variable of central interest is `MonthGive`, which is

Figure 5.8: Recode Variables Dialog

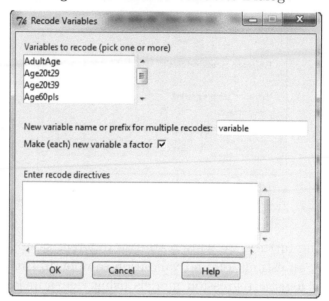

a binary factor that takes on the level "Yes" if a donor signed up for the monthly giver program and the level "No" if a donor did not. **View data set** to confirm these values. We will create a new numeric variable that takes on the value 1 if the level of MonthGive is "Yes" and the value 0 when the level of MonthGive is "No." The tricky and useful bit is that the average of this variable that consists of 0s and 1s will be between 0 and 1, and is the proportion of donors who joined the monthly giver program. (To convince yourself, calculate by hand the average for ten donors, three of whom are monthly givers). This allows simple and easy-to-understand visualizations. To create the new numeric MonthGive.Num variable use the pull-down menu option **Data → Manipulate variables → Recode variable...**, which will bring up the dialog box shown in Figure 5.8.

4.

In the **Recode Variable** dialog box, **select MonthGive** as the "Variable to recode (pick one or more)"; enter MonthGive.Num in the "New variable name or prefix" field. We want the new variable to be numeric so we can take averages, so **UNcheck the "Make new variable a factor" box**. In the expression box (the entry box located on the lower right-hand side of the Recode Variable dialog box), **enter "Yes" = 1, press <Return>, and on the second line enter "No" = 0**. When you are done, your **Recode Variable** dialog box should look like the one shown in Figure 5.9. Remember that R is case sensitive, so the entries need to be exactly as shown.

Figure 5.9: The Completed Recode Variables Dialog

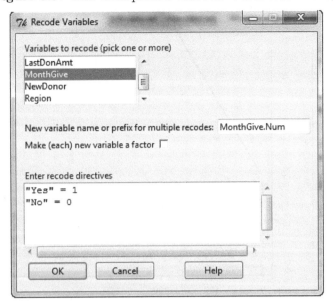

One important point that needs to be made is that the lift chart tools we present in the next chapter only work if the target variable is a two-level (binary) *factor*. While we will use `MonthGive.Num` for visualization and exploration, it is not a factor, therefore you will want to use the factor `MonthGive` as the target variable when you move on to constructing your logistic regression models. You have to remember this since the logistic regression tools are perfectly happy using either a numeric variable with zeros and ones, or a factor variable with "Yes" and "No" as the target variable.

5.

Press **OK** in the Recode Variable dialog box to create `MonthGive.Num`, and then create a scatterplot (through the pull-down menu option **Explore and Test → Visualize→ Scatterplot. . .**) using `MonthGive.Num` as the *y*-variable and `AveDonAmt` as the *x*-variable. You should have a plot that looks like the one shown in Figure 5.10.

6.

The circles represent the data points. The estimated "smooth line" (solid line) attempts to fit the data points in Figure 5.10, and indicates there is a strong nonlinear (in this case, concave) relationship between `AveDonAmt` and `MonthGive`. The concave relationship suggests that a logarithm transformation of `AveDonAmt` may be in order for this variable. While the smooth line in

Figure 5.10: Monthly Giver vs. Average Donation Amount

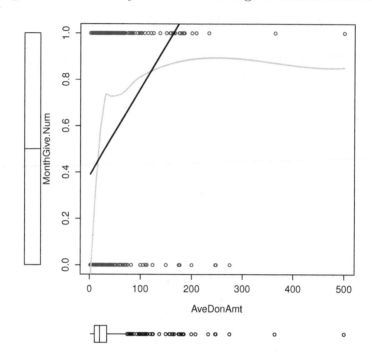

the scatterplot is one of the easiest ways to look for the nonlinear relationships between two variables, it tends to run into problems in binary (0/1) data when there are very unequal numbers of the two binary responses. Fortunately, we have oversampled the Yes responses (the 1s) so that there is a 50–50 split between Yes and No values, which allows the smooth line to detect the nonlinear effect of `AveDonAmt`. However, if we had not oversampled the Yes values, almost all of the variables would have a *y*-value of 0, and the smooth line would have been a straight line at 0 for `MonthGive.Num`.

Fortunately, there is a second method that allows us to look for nonlinear relationships in cases where the oversampling is not as extensive (i.e., when the sample contains fewer than 50% positive responses to the target variable). The second approach involves binning (which we have done before) the continuous predictor variable `AveDonAmt` into categories, and then using a plot of means to visualize the relationship between the target and predictor variables. This method works best if there is a fairly even distribution of the predictor variable across its range of values. Things are a bit less ideal when there is a large skew (e.g., when a large number of customers have a low value, while a small number have a high value) in the predictor of interest, but even in these cases useful information can be obtained. Use the pull-down menu option **Data →** **Manipulate variables → Bin a numeric variable. . .** to create the new variable

Figure 5.11: Plot of Means Dialog

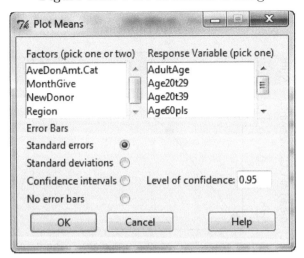

AveDonAmt.Cat. Specifically, **bin AveDonAmt into four equal-count bins and select "ranges" as the level names.** As usual, check the result with **View data set.** AveDonAmt turns out to be highly skewed, which is why we are using equal count bins. If it was less skewed, we would have selected equal interval bins, which allows for a plot of means that is a bit easier to interpret. However, equal interval bins with highly skewed variables can end up with bins that have no members, which causes problems. Next, use the pull-down menu option **Explore and Test →Visualize → Plot of means. . .** to bring up the dialog box shown in Figure 5.11.

7.

In the Plot Means dialog box **select AveDonAmt.Cat as the "Factors (pick one or two)"** and **MonthGive.Num as the "Response Variable (pick one)."** While not required, you may want to select "No error bars" as the "Error Bars" option since if there are two or fewer observations in a particular bin, error bars cannot be created, and the plot of means tool will fail. Click **OK** and the plot in Figure 5.12 will be created.

8.

You can see the ranges represented by each level of AveDonAmt.Cat by looking at the labels along the horizontal axis of the figure. For example, the second label, (10,20.6], indicates the range of donations in the second level is from \$10 (the round left bracket indicates "not including \$10") to \$20.60 (the square right bracket indicates "including \$20.60"). Recall that the mean of MonthGive.Num is the proportion of "1" or "Yes" values in each category.

Figure 5.12: Plot of Means of Monthly Giver vs. Average Donation Amount

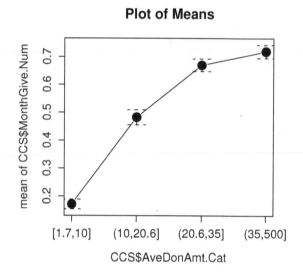

Slightly less than half of the individuals in the second donation category are members of the program. Donors with low average donation amounts have fewer members in the monthly giver program, and we can see that the increase with larger donations is of the concave (or decreasing-returns) structure, rather than linear.

9.

The plot of means is a reasonable way of visualizing the effect of a continuous predictor variable (after binning) on a binary target variable, and it is the first choice in viewing the effect of a factor (nominal) predictor variable on the target variable. Another way would be to use a boxplot of the continuous variable, grouped by the two levels of the binary target variable, although this would not give as much detailed insight.

10.

Create a plot of means using `Region` as the factor and `MonthGive.Num` as the response variable. Once you have done this, you should have a plot like the one shown in Figure 5.13. Figure 5.13 suggests that later on we may want to consider reducing the number of levels of Region to two or three. Specifically, we may want to create a new region variable that combines regions R1 (Vancouver Island), R2 (Greater Vancouver), and R3 (the Fraser Valley) into a single region, and R4 (the North Coast of BC), R5 (South and Central Interior of BC), and R6 (the Northern Interior of BC and the Yukon) into a

Figure 5.13: Monthly Giver vs. Region Plot of Means

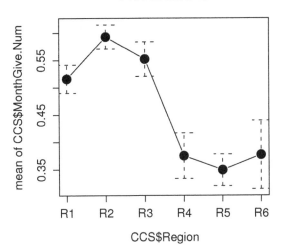

second region. In the interest of speculating (or hypothesizing, or theorizing, or storytelling), we note that a possible causal explanation of what is driving these apparent region effects is related to how urbanized a region is, with donors living in more urban regions being more likely to join the monthly giving program.

11.

Exercise: Ensure you have mastered the above material before continuing on by examining the relationship between `MonthGive.Num` and other potential predictor variables. Use the help file on the CCS data set to see the variable definitions and get some idea of what might be interesting. At a minimum, look at `LastDonAmt`, `DonPerYear`, and `YearsGive` as predictors of membership in the Monthly Giving Program, and identify any that appear to be nonlinear.

12.

We are now ready to begin building a logistic regression model. However, before doing this we need to consider which variables are likely candidates for predictors. This stage makes use of managerial knowledge, and is essentially creating some theory to guide our modeling. This will hopefully speed up the development of a *reasonably good* — if not the absolutely best — model. Based on discussions with CCS personnel, a number of donor characteristics were identified as generally being related to donation behavior on the part of donors. These characteristics include measures related to past donation behavior (captured by the variables `AveDonAmt`, `LastDonAmt`, `DonPerYear`, and `YearsGive`) of the donor, the region in which the donor resided, and (anecdotally) his or her demographic profile. The CCS had never surveyed its donors to develop a demographic profile to determine if and how they differed from the general population. However, donors were thought to be on average older, live in households with fewer members present, and to have higher incomes than the general population. In addition, there was some evidence that some ethnic groups (recent Chinese immigrants in particular) were more likely to be donors. One issue with income as a predictor variable is that an individual's income tends to be highly correlated with his or her level of educational attainment. As a result, a decision was made to examine the potential effect of educational attainment as well.

13.

Although the CCS did not have demographic information for individual donors, it did have the postal code of each donor, and access to neighborhood-level census data. The postal code was used to assign the neighborhood-level (typically 300 to 400 households) average demographic characteristics in which the individual resided, based on the 1996 Census of Population, to the individual. These geo-demographic variables are therefore not true measures of the demographic characteristics of the individual, but are proxy measures based on population averages of the small geographic area in which a donor resides. For instance, we do not know the donor's actual income, but we can determine the average income level among households that reside in the small geographic area in which the donor resides, and use this as a proxy measure for that individual's income level. We do need to remember that these variables are only proxies of the neighborhood when *interpreting* our estimated model.[4] A number of variables related to the age distribution of the donors' neighborhoods were included in the data set to capture the possible effects of donor age, but we will only consider two of these: `Age20t29` (the percentage of people residing in the area who are between 20 and 29 years of age) and `Age70pls` (the percentage of people residing in the area who are 70 years of age or older). Two variables related to household size are included in the data set, but we will only examine the effect of `hh1mem` (the percentage of households with only a single member present). The effect of ethnicity and recent immigration is measured using the variable `EngPrmLang` (the percentage of households in the area in which the donor resides where English is the primary language spoken at home). Income is measured using the variable `AveIncEA` (the average annual income level of households residing in the same

---

[4]Technically, assuming that the area average represents the value for a household is known as an ecological fallacy.

Figure 5.14: The Generalized Linear Model Dialog

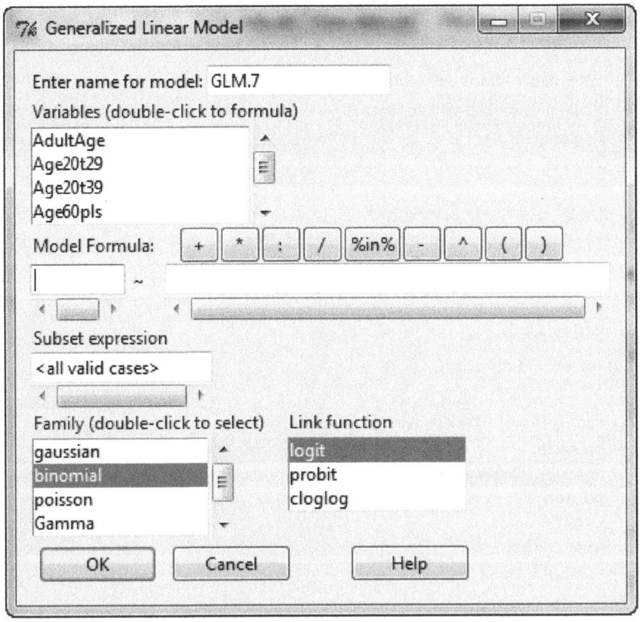

area as the donor). Finally, `FinUnivP` (the percentage of individuals in the area who have obtained a university degree) is used to capture the potential effect of educational attainment.

14.

To actually estimate the logistic regression model use the pull-down menu option **Models → Statistical models → Generalized linear model...**, which brings up the dialog box shown in Figure 5.14.

15.

The **Generalized Linear Model** dialog box is very similar to the **Linear Model** dialog box we saw in the linear regression tutorial. What is different are the "Family (double-click to select)" and "Link function" fields at the bottom of the **Generalized Linear Model** dialog box. Luckily, the default selections in these two fields correspond to the logistic regression model, so we do not need to alter them. The remaining fields are identical (and behave identically) to those in the **Linear Model** dialog box. To indicate that we have not transformed any variables yet, **enter LinearCCS in the "Enter name for model:" field**. The target variable is **MonthGive,** the categorical factor variable. Our continuous MonthGive.Num would work equally well, but later on we will be using lift charts for model assessment, which require binary factor variables as targets.

Figure 5.15: The Completed Generalized Linear Model Dialog

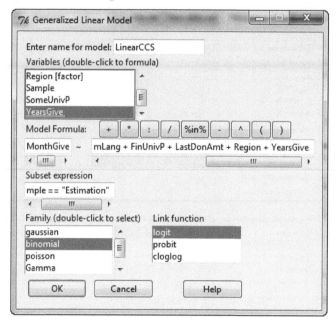

MonthGive should be on the left-hand side of the "Model Formula:"
field, while the variables Age20t29, Age70pls, AveDonAmt, AveIncEA,
DonPerYear, EngPrmLang, FinUnivP, LastDonAmt, Region **and** YearsGive
(as with a linear model formula, the predictor variable names will be separated
by a "+"). The one new field we will be working with is "Subset expression,"
which is the way in which we implement the use of different samples. To
estimate the model using only records from the Estimation sample, **enter**
into this field the expression Sample == "Estimation." The use of the
two equal signs is intentional, and indicates a *test* for equality. If the statement
is true, that is, if the variable Sample has the value Estimation, the record
is selected for analysis. The single equal sign that we have used in previous
situations is known as an assignment operator, and assigns a new value to a
variable. Once you have done all of this, your dialog box should look like the
one shown in Figure 5.15. When it does, press **OK** to estimate the model,
the summary of which will appear in the R Commander output window, and
should look "approximately" (e.g., the results may have a different variable
order) like those shown in Figure 5.16.

16.

Figure 5.16 indicates that variables related to past donation behavior
(DonPerYear and LastDonYear) have the greatest impact statistically, while
the only demographic variables that seem to have any impact is AveIncEA

Figure 5.16: LinearCCS Model Results

```
Output Window                                                          Submit

-3.0935  -0.9952  -0.6402   1.0908   1.8571

Coefficients:
                Estimate Std. Error z value Pr(>|z|)
(Intercept)   -1.201e+00  8.489e-01  -1.415 0.156993
Age20t29       6.572e-01  1.907e+00   0.345 0.730357
Age70pls       1.529e-01  9.743e-01   0.157 0.875263
AveDonAmt     -6.008e-03  6.837e-03  -0.879 0.379543
AveIncEA      -9.636e-06  6.107e-06  -1.578 0.114563
DonPerYear     1.036e+00  2.060e-01   5.030 4.9e-07 ***
EngPrmLang     1.238e-01  6.980e-01   0.177 0.859249
FinUnivP       4.318e-01  1.034e+00   0.418 0.676199
LastDonAmt     2.303e-02  6.829e-03   3.372 0.000746 ***
Region[T.R2]   4.927e-01  2.573e-01   1.915 0.055481 .
Region[T.R3]   4.503e-01  2.537e-01   1.775 0.075928 .
Region[T.R4]  -1.268e-01  3.426e-01  -0.370 0.711254
Region[T.R5]  -2.913e-01  2.485e-01  -1.172 0.241043
Region[T.R6]  -4.426e-01  4.376e-01  -1.011 0.311877
YearsGive      4.099e-02  2.717e-02   1.509 0.131415
---
Signif. codes:  0 '***' 0.001 '**' 0.01 '*' 0.05 '.' 0.1 ' ' 1

(Dispersion parameter for binomial family taken to be 1)

    Null deviance: 1108.99  on 799  degrees of freedom
Residual deviance:  982.63  on 785  degrees of freedom
AIC: 1012.6

Number of Fisher Scoring iterations: 5


> 1 - (LinearCCS$deviance/LinearCCS$null.deviance) # McFadden R2
[1] 0.1139388
```

(but still isn't quite statistically significant). The McFadden R^2 is a statistic that ranges between 0 and 1, with 1 being a perfect fit, and can be interpreted similarly to R^2 in linear regression. However, McFadden R^2 values are typically less than R^2 values, so we tolerate lower values for McFadden R^2. For the model the McFadden R^2 is a low, but acceptable, 0.114. The AIC value, 1012.6, is used like adjusted R^2 for choosing a set of predictors for a model by comparing its value across different models, since it includes both the goodness of fit and a penalty for using more variables. A reasonable critierion to select a model is to choose the one that minimizes the AIC.

Finally, it is difficult to determine if **Region** has any effect. One of the reasons is that R1 is the "base case" region, which Figure 5.13 indicates is in the high group, but has the lowest average in this group. Recall that the statistics for nominal variables are measuring whether each region has a significantly different impact from R1 (whereas for the continuous variables, the measure is whether the coefficient is significantly different from zero). The difference between the highest and lowest values in Figure 5.13 has the best chance of actually being significant. As with a linear regression model, a better way to examine if Region has an influence on its own is to use the pull-down menu option **Assess** → **Hypothesis tests** → **ANOVA** table, which produces the results in the R Commander output window shown in Figure 5.17. The results in this figure indicates that, on the whole, Region does have a statistically

Figure 5.17: LinearCCS ANOVA Results

```
> Anova(LinearCCS, type="II", test="LR")
Analysis of Deviance Table (Type II tests)

Response: MonthGive.Num
            LR Chisq Df Pr(>Chisq)
Age20t29      0.1189  1  0.7301842
Age70pls      0.0246  1  0.8752618
AveDonAmt     0.7661  1  0.3814137
AveIncEA      2.5569  1  0.1098129
DonPerYear   30.0458  1   4.22e-08  ***
EngPrmLang    0.0314  1  0.8593249
FinUnivP      0.1747  1  0.6759754
LastDonAmt   12.5129  1  0.0004042  ***
Region       12.4327  5  0.0293165  *
YearsGive     2.2941  1  0.1298686
---
Signif. codes:  0 '***' 0.001 '**' 0.01 '*' 0.05 '.' 0.1 ' ' 1
```

significant effect, although a fairly weak one. Based on this, and Figure 5.13, we probably should reduce the number of levels in this variable and recalibrate the model.

17.

One thing that is surprising is that there appears to be no relationship between AveDonAmt and being a member of the monthly giving program. However, Figure 5.10 indicates that the influence of this variable is strongly non-linear (concave) in nature. Thus, a logarithm transformation seems to be in order. It turns out that the logarithm of an average (which is what we have for all our demographic measures in the CCS database) is problematic in terms of its interpretation (i.e., the log of the mean does not equal the mean of the log transformed original variable). As a result, we will only transform those variables that relate directly to an individual donor (i.e., AveDonAmt, LastDonAmt, DonPerYear, and YearsGive). At this point we will encounter a common technical concern with log transformations: the natural logarithm of zero (as well as negative numbers) is undefined, so if the variable we wish to transform has zeros, the computation will fail. Before doing log transformations, we must always determine which variables have zero values, and modify those variables. This is not an issue for AveDonAmt and LastDonAmt since to be a donor, an individual had to have given a non-zero amount in the past. DonPerYear and YearsGive could have zero values (years of giving to the CCS is measured as integer full years, so recent donors could still be assigned 0). You can easily determine if a variable has zero or negative values by using the pull-down menu option **Explore and Test → Summarize → Numerical summaries. . . .** The dialog box that then appears allows you to select a variable for which summary statistics will be printed to the R Commander output window. **Select**

Figure 5.18: Numerical Summaries for DonPerYear and YearsGive

```
> numSummary(CCS[,c("DonPerYear", "YearsGive")], statistics=c("mear
                mean        sd  0%        25% 50% 75% 100%    n
DonPerYear 0.6157976 0.5008012   0 0.2727273 0.5   1    5 1600
YearsGive  4.8593750 3.1716931   0 2.0000000 4.0   8   14 1600
```

YearsGive, and **DonPerYear**. When you are done, the summary statistics for these two variables should appear as they do in Figure 5.18. In this instance, the statistic of interest is the 0% quantile (which gives the minimum value of the variable). As can be seen in Figure 5.18, the minimum value of both **DonPerYear** and **YearsGive** is zero.

18.

For **DonPerYear** and **YearsGive** a modification of the log transformation is needed. Specifically, for variables with zero (but no negative) values, the transformation $\log(X + 1)$ will eliminate the problem with zeros, where X is the original variable to be transformed. As in the linear regression tutorial, use the pull-down menu option **Data → Manipulate variables → Compute new variable...** and this transformation to create new log transformed variables Log.AveDonAmt, Log.LastDonAmt, Log.DonPerYear, and Log.YearsGive. For AveDonAmt and LastDonAmt a simple log transformation, $\log(X)$, should be used. Check that the variables have been created and that they have no missing values (**View Data set**).

19.

Run a second logistic regression model, this time **entering LogCCS in the "Enter name for model:" field**. The target variable is again MonthGive and should be on the left-hand side of the "Model Formula:" field, while the variables Log.AveDonAmt, Log.DonPerYear, Log.LastDonAmt, Log.YearsGive, Region, Age20t29, Age70pls, EngPrmLang, AveIncEA, FinUnivP, and should be the right-hand side variables. Remember to **set the Sample=="Estimation."** Your results for this second model (with the exception of possible variable order differences) should be identical to those shown in Figure 5.19. The overall fit of the model has improved considerably, with the McFadden R^2 rising to a much more reasonable 0.177 and the AIC falling (improving) to 942. The log transformed version of **AveDonAmt** is now barely statistically significant (only at the 10 percent level, not 5 percent). However, the log version of **YearsGive** is not statistically significant, nor are any of the demographic variables. As before, determining whether or not **Region** is significant is difficult from this output, but can easily be de-

Figure 5.19: LogCCS Model Results

```
Output Window

Coefficients:
                 Estimate Std. Error z value Pr(>|z|)
(Intercept)    -3.691e+00 9.487e-01  -3.890  0.00010 ***
Log.AveDonAmt   4.695e-01 2.600e-01   1.806  0.07097 .
Log.DonPerYear  1.111e+00 3.420e-01   3.248  0.00116 **
Log.LastDonAmt  6.022e-01 2.371e-01   2.540  0.01109 *
Log.YearGive   -1.656e-01 1.390e-01  -1.192  0.23345
Region[T.R2]    5.062e-01 2.698e-01   1.876  0.06065 .
Region[T.R3]    4.360e-01 2.658e-01   1.640  0.10098
Region[T.R4]   -1.884e-01 3.593e-01  -0.524  0.59998
Region[T.R5]   -1.708e-01 2.620e-01  -0.652  0.51448
Region[T.R6]   -4.604e-01 4.617e-01  -0.997  0.31868
Age20t29       -5.209e-01 1.994e+00  -0.261  0.79394
Age70pls        3.589e-01 1.021e+00   0.351  0.72531
EngPrmLang      6.230e-01 7.220e-01   0.863  0.38823
AveIncEA       -6.310e-06 6.359e-06  -0.992  0.32105
FinUnivP       -4.381e-01 1.071e+00  -0.409  0.68258
---
Signif. codes:  0 '***' 0.001 '**' 0.01 '*' 0.05 '.' 0.1 ' ' 1

(Dispersion parameter for binomial family taken to be 1)

    Null deviance: 1108.99  on 799  degrees of freedom
Residual deviance:  912.33  on 785  degrees of freedom
AIC: 942.33

Number of Fisher Scoring iterations: 4


> 1 - (LogCCS$deviance/LogCCS$null.deviance) # McFadden R2
[1] 0.1773371
```

termined using the pull-down menu option **Assess → Hypothesis tests →
ANOVA table** (Figure 5.20), which indicates it is, but only at 0.10 percent
level.

20.

Comparing the results of the two logistic regression models indicates that a
log transformation of some of the variables noticeably improves the overall
fit of the model (lower AIC and higher McFadden R^2) but, based on changes
in variable significance levels, it may well be the case that not all variables
should be log transformed. In addition, some of the variables we included in
the original model seem to have little or no explanatory power, and could be
removed, while (based on Figures 5.13 and 5.20) the Region variable would
likely benefit from a reduction in the number of levels. To help make an assessment of which variables should be log transformed and which variables
should be removed, Table 5.1 provides a comparison of the absolute values of
$Pr(>|z|)$ of all predictor variables except `Region` for both the `LinearCCS` and
`LogCCS` models.

Figure 5.20: LogCCS ANOVA Results

```
> Anova(LogCCS, type="II", test="LR")
Analysis of Deviance Table (Type II tests)

Response: MonthGive.Num
               LR Chisq Df Pr(>Chisq)
Log.AveDonAmt    3.2125  1  0.0730756 .
Log.DonPerYear  10.8590  1  0.0009832 ***
Log.LastDonAmt   6.7393  1  0.0094312 **
Log.YearGive     1.4196  1  0.2334625
Region           9.9016  5  0.0780728 .
Age20t29         0.0681  1  0.7940791
Age70pls         0.1236  1  0.7252020
EngPrmLang       0.7393  1  0.3898763
AveIncEA         0.9986  1  0.3176492
FinUnivP         0.1670  1  0.6827739
---
Signif. codes:  0 '***' 0.001 '**' 0.01 '*' 0.05 '.' 0.1 ' ' 1
```

Table 5.1: LinearCCS and LogCCS Variable p-Values

Variable	LinearCSS	LogCCS
AveDonAmt*	0.380	0.070
LastDonAmt*	0.000	0.011
DonPerYear*	0.000	0.001
YearsGive*	0.131	0.233
Age20t29	0.730	0.794
Age70Pls	0.875	0.725
AveIncEA	0.115	0.321
EngPrmLang	0.859	0.388
FinUnivP	0.676	0.683

21.

An examination of Table 5.1 reveals that the only variable which the log transformation improved was `AveDonAmt`, while the demographic variables `Age20t29`, `Age70pls`, `EngPrmLang`, and `FinUnivP` have low z-statistics in both models. Based on the table, all demographic variables could be removed. Although, `AveIncEA` is nearly significant, so we will keep it in the set of variables for some additional testing. We will also keep `YearsGive` for the same reason, and only use a logarithmic transformation on `AveDonAmt`. In addition, as argued earlier, `Region` should be recoded to have only two levels (one containing original regions R2, and R3, and the other containing original regions R1, R4, R5, and R6). The reason for this choice is that regions R2 and R3 were found to be statistically different from R1 in both estimated models, but this is not true for regions R4, R5, and R6. Use the pull-down menu option **Data** → **Manipulate variables** → **Relabel factor levels...** to create the variable `New.Region` in which regions R2, and R3 are assigned to the level "VanFraser" (since R2 is Greater Vancouver, while R3 is the Fraser River Valley), and regions R1, R4, R5, and R6 are assigned to the level "Other." View your data to check you have the new variable with correct values. Following this, **estimate a third logistic regression model (which you should name MixedCCS), which has AveIncEA, DonPerYear, LastDonAmt, Log.AveDonAmt, New.Region and YearsGive** as predictor variables. Your results for this model should correspond to those contained in Figure 5.21.

22.

Figure 5.21 indicates that this model (based on both the McFadden R^2 and AIC) fits much better than did the original `LinearCCS` model. The McFadden R^2 is lower (indicating a loss of fit) for this model than for the `LogCCS` model (although the AIC favors the new model over the `LogCCS` model due to the reduction in the number of estimated parameters). The individual z-statistics indicate that all variables except `YearsGive` are statistically significant at the 10 percent level or better. At this point it seems appropriate to give up on `YearsGive` as a predictor variable. One of the most important changes is that `LastDonAmt` is now only significant at the 10 percent level, which is surprising. However, both a scatterplot and a plot of means for `LastDonAmt` indicate that the relationship between this variable and being a member of the Monthly Giving Program is concave in nature. As a result, **estimate one final logistic regression model, which you should label MixedCCS2, which uses AveIncEA, DonPerYear, Log.AveDonAmt, Log.LastDonAmt, and New.Region as predictor variables.** Your results for this model should be identical to those of Figure 5.22.

Figure 5.21: MixedCCS Model Results

```
Output Window

Coefficients:
                          Estimate  Std. Error  z value  Pr(>|z|)
(Intercept)              -3.852e+00  4.701e-01   -8.195  2.50e-16 ***
AveIncEA                 -7.694e-06  4.078e-06   -1.887  0.05921 .
DonPerYear                8.761e-01  2.088e-01    4.197  2.71e-05 ***
LastDonAmt               -5.158e-03  2.724e-03   -1.893  0.05833 .
Log.AveDonAmt             1.193e+00  1.407e-01    8.474  < 2e-16 ***
New.Region[T.VanFraser]   4.797e-01  1.677e-01    2.860  0.00424 **
YearsGive                 2.164e-02  2.762e-02    0.784  0.43332
---
Signif. codes:  0 '***' 0.001 '**' 0.01 '*' 0.05 '.' 0.1 ' ' 1

(Dispersion parameter for binomial family taken to be 1)

    Null deviance: 1108.99  on 799  degrees of freedom
Residual deviance:  915.42  on 793  degrees of freedom
AIC: 929.42

Number of Fisher Scoring iterations: 4


> 1 - (MixedCCS$deviance/MixedCCS$null.deviance) # McFadden R2
[1] 0.1745503
```

Figure 5.22: MixedCCS2 Model Results

```
Output Window

Coefficients:
                          Estimate  Std. Error  z value  Pr(>|z|)
(Intercept)              -3.384e+00  4.024e-01   -8.408  < 2e-16 ***
AveIncEA                 -7.442e-06  4.132e-06   -1.801  0.07169 .
DonPerYear                8.089e-01  1.864e-01    4.340  1.43e-05 ***
Log.AveDonAmt             4.166e-01  2.581e-01    1.614  0.10653
New.Region[T.VanFraser]   4.792e-01  1.676e-01    2.859  0.00426 **
Log.LastDonAmt            6.214e-01  2.375e-01    2.617  0.00887 **
---
Signif. codes:  0 '***' 0.001 '**' 0.01 '*' 0.05 '.' 0.1 ' ' 1

(Dispersion parameter for binomial family taken to be 1)

    Null deviance: 1108.99  on 799  degrees of freedom
Residual deviance:  911.96  on 794  degrees of freedom
AIC: 923.96

Number of Fisher Scoring iterations: 4


> 1 - (MixedCCS2$deviance/MixedCCS2$null.deviance) # McFadden R2
[1] 0.1776685
```

Note: A feature of the software is that the dialog box will start with the parameters (with the exception of the model name) of whichever model is selected in the blue Model button. This allows you to generate new models quickly by starting with an existing model that is close to the new one.

23.

Figure 5.22 indicates that this model fits the data better than any of the other three models we have estimated (two of which had many more estimated parameters) based on both the McFadden R^2 and AIC statistics, and that all the predictor variables except `AveDonAmt` are statistically significant (it turns out that removing `AveDonAmt` significantly worsens both the AIC and McFadden R^2). The problems we experienced in obtaining significant results for both `AveDonAmt` and `LastDonAmt` are related to the fact that these variables are highly correlated with one another. In the context of linear regression, this is known as collinearity, and the effect in logistic regression (and other models we will encounter later) is similar. It appears that both variables help in predicting monthly giving program membership. If our purpose is simply to apply the model for prediction this is not a major problem. If we wish to determine which variables matter on the basis of significance levels, however, correlated variables are problematic. Correlation causes the reported estimates of the standard errors on the coefficients of the correlated variables to be inflated above their true values, which can make variables that truly matter appear to be statistically insignificant.

24.

Exercise: In two or three sentences, describe how the members of the monthly giving program differs from nonmembers.

25.

In the tutorial for the next chapter, we will be comparing the four models you estimated in this tutorial using lift charts, a technique that only considers the predictive abilities of the models. In subsequent tutorials, we will also want to have the last model (`MixedCCS2`) available. To make doing these tutorials easier, save the workspace files from this tutorial so that the four models are readily available to you when you start the next tutorial. Use **File → Save R Workspace as. . .** to bring up the dialog box, and **save the file as CCSLogistic.RData**. Remember that the extension ".RData" is case sensitive.

Chapter 6

Lift Charts

Lift charts (Barry and Linoff, 1997) are a graphical tool that both indicate the relative predictive power of different possible candidate models, and allow managers to quickly apply an estimated predictive model in order to address specific managerial questions.

This chapter describes how lift charts are created and how they are used to profitably target customers in direct marketing applications. To make the description of how lift charts are created concrete, this note makes use of an example based on a random sample of 50 donors from the Canadian Charitable Society data set (CCS in the BCA data library) and the `MixedCCS2` logistic regression model that is developed in the logistic regression tutorial in Chapter 5. The managerial use of lift charts is illustrated using a second example involving a regional telephone service provider's sales campaign for DSL service.

While the example approach makes the concepts behind the creation of lift charts more concrete, the disadvantage is that lift chart methods rely on there being many customers in the database, breaking those customers into a set of groups (traditionally ten groups), and then presenting averages for each of these groups. The use of only 50 observations allows us to reasonably look at all the data, but makes dividing the data into ten separate groups impractical for reasons of random sampling error. Consequently, instead of using ten groups in this note, the data will be divided into five groups. However, the concepts presented are applicable to an arbitrary number of groups.

6.1 Constructing Lift Charts

6.1.1 Predict, Sort, and Compare to Actual Behavior

Table 6.1 presents the data for the 50 donors randomly chosen from the CCS data set. The first column indicates whether the donor actually joined the monthly giver program, and the second column is the estimated probability (calculated using the MixedCCS2 logistic regression model) that they would join the monthly giver program based only on their past donation behavior, the region of BC and the Yukon in which the donor resides, and the demographic characteristics of people living in the small geographic area in which the donor resides. The 50 observations have then been sorted from the highest to the lowest predicted probability. In this case, the predicted probability is a score that is given to each customer. The process of scoring a customer database, by this or any of a number of other methods, is a common method of quantifying our assessment of better and worse customers. The score becomes a targeting variable that allows us to approach more desirable customers.

Our ultimate goal is to apply a model to customers who have never joined the monthly giver program—that is, customers for whom we have not yet observed the "yes" and "no" information. Before doing that, we really should assess how well the model predicts. Here, "good" means how much better we can target the individuals in the database by using the model than by a purely random selection approach. The best way to assess the model is to compare the model predictions with already known behavior of individuals. Specifically, the individuals for whom we have "yes" and "no" data.

Table 6.1 has been divided into five groups of ten. In the first group eight out of ten donors (80 percent) joined the monthly giver program, while five out of ten (50 percent) joined the monthly giver program in the second group, only three out of ten (30 percent) joined the monthly giver program in the third group, and only two of ten (20%) in the final two groups. The group membership, based only on predicted behavior, is strongly related to actual behavior, exactly what we would hope. The prediction is not perfect, but much better than we could do without the model. Without the model, any random selection of 100 individuals from this database would give, on average, 40 members of the program, so any procedure that will allow us to find donors with a probability of joining greater than 40% is an improvement.

In this chapter, we will discuss how to construct two of the most commonly used (and managerially relevant) lift chart types: the "incremental response rate" chart and the "total cumulative response" chart. The incremental chart is a plot of the information described in the preceding paragraph, and a plot of this information can be found in Figure 6.1.

In Figure 6.1, the horizontal axis in the plot displays the cumulative percentage of the sample each group represents. For managerial purposes, we read this axis by saying, "Suppose we contacted the top 20% (second 20%, third 20%, etc.) of our database... ". The data points for each group are the diamonds in the plot, and are connected together by solid lines. They allow our manager to finish the above statement as "... then we would have an incremental response rate in that quintile of 80% (50%, 30%, etc.)."

The horizontal line at the incremental response rate of 40% is the baseline

Table 6.1: A Random Sample of 50 Donors from the CCS Database Sorted by Fitted Probability

Sorted Rank	Monthly Giver?	Fitted Probability	Sorted Rank	Monthly Giver?	Fitted Probability
1	Yes	0.976	31	Yes	0.410
2	Yes	0.934	32	No	0.408
3	Yes	0.897	33	No	0.407
4	Yes	0.894	34	No	0.386
5	No	0.890	35	Yes	0.352
6	Yes	0.877	36	No	0.345
7	No	0.865	37	No	0.339
8	Yes	0.832	38	No	0.327
9	Yes	0.805	39	No	0.324
10	Yes	0.778	40	No	0.288
11	Yes	0.749	41	Yes	0.265
12	No	0.744	42	Yes	0.245
13	No	0.710	43	No	0.233
14	No	0.650	44	No	0.205
15	Yes	0.648	45	No	0.197
16	Yes	0.646	46	No	0.190
17	Yes	0.645	47	No	0.181
18	No	0.644	48	No	0.172
19	No	0.626	49	No	0.156
20	Yes	0.611	50	No	0.097
21	No	0.592			
22	No	0.565			
23	Yes	0.543			
24	Yes	0.537			
25	No	0.519			
26	Yes	0.514			
27	No	0.492			
28	No	0.471			
29	No	0.440			
30	No	0.435			

Figure 6.1: The Incremental Reponse Rate Chart for the Sample

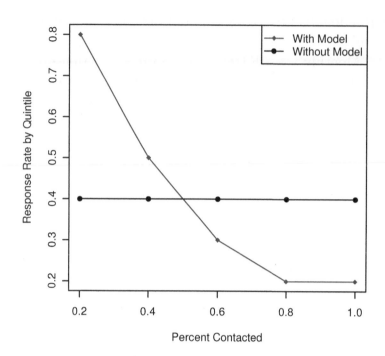

from which we assess model lift. It corresponds to the expected response rate we would achieve if we randomly selected donors from the sample (i.e., a 40 percent response rate).

In a standard incremental response chart the sample would have been divided into ten groups (or deciles) and not quintiles as done here. The problem with using deciles for this example is that there would be only five donors in each group, and such a small number can result in very jagged lift charts. To illustrate this point, an incremental response rate chart based on decile groups for the sorted sample of 50 donors is presented in Figure 6.2. In general, the more observations in each group, the smoother the resulting lift chart is likely to be.

The other useful lift chart, the total cumulative response chart, is constructed by taking the sum of total positive responses (donors who joined the monthly giver program) up to and including a particular group, and dividing that sum by the total number of positive responses for the sample. For the random sample of 50 donors, the total number of "positive responders" (i.e., monthly givers) is 20, and the number of these that are in the first group (the "best" fifth of the sample) is eight. Thus, if we had contacted the top 20% of the group as identified only by the model, we would have captured 8 of the 20 true positives, or 40%. If we had contacted the top 40%, we would have captured 13 of the 20 givers, or 65%. The total cumulative captured response over

Figure 6.2: The Incremental Response Rate Chart Using Deciles

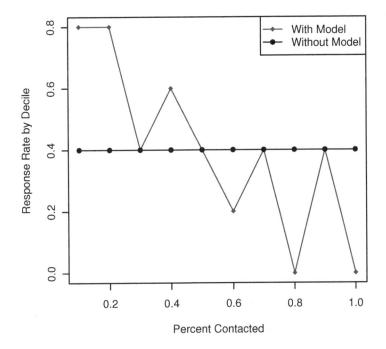

all five groups is, of course, 100%—if we contact everyone, we capture all responders. The cumulative response chart always ends at 100%. Figure 6.3 provides a plot of the total cumulative response chart. In the plot the 45-degree line is the baseline from which to assess model lift. It corresponds to the expected total cumulative response when donors are selected at random from the database. Generally, the more rapidly the total cumulative response approaches 1, the greater is the potential benefit from using the predictive model to target customers.

6.1.2 Correcting Lift Charts for Oversampling

So far we have been implicitly assuming that the proportion of members of the monthly giver program in the sample used for model estimation is also true of the entire Canadian Charitable Society donor base. As we know, however, the members have been oversampled here to allow the model to detect predictors of membership. Since our actual campaign will use the complete database, we will need to correct for this oversampling to determine the lift and true response rates we can expect when we apply our model to the original database. In the original complete database members represent only 1% of donors, not 40% as in the previous calculations.

The easiest way to think about this correction is to imagine a new propor-

Figure 6.3: The Total Cumulative Response Rate Chart for the Sample

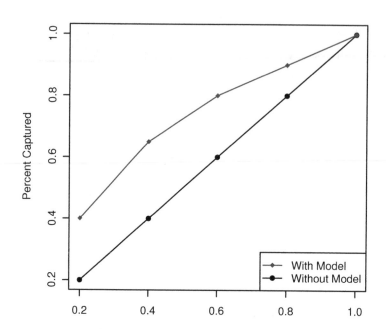

tionate sampling of the original database that continued to draw donors until we had 20 donors who opted into the Monthly Giver Program. To get 20 donors that are in the program at a 1% response rate, we would expect to have to draw 2000 donors from the original database—20 members, and 1980 nonmembers. Comparing this to our oversampled set of 20 members and 30 nonmembers, we see that each one of the 30 nonmembers has to stand in for 66 nonmembers in the complete database (1980 / 30 = 66). We now simply reconstruct Table 6.1.1 replacing each nonmember with 66 nonmembers.

Table 6.2 presents the results giving each nonmember a weight of 66. An additional column is included to track the new percentages so that we can redefine our quintiles. The lines show the cutoff points at each quintile (these are not exactly at the quintile points but as close as we can get—with more records one can get very close). Table 6.2 can now be used to construct the new lift charts, which will tell us what to expect in the way of lift and response when we apply our model to the entire database.

As a quick check on the new lift chart, note that a 40% overall rate in the sample must become a 1% overall rate in the database. This means that the incremental response rates will be roughly 40 times smaller. It won't be exactly 40 times, because the cutoff points for the five bins will change. Since there are fewer "no's" in the earlier part of the database than the later, our correction adds fewer new individuals to the earlier part of the database than the later.

Table 6.2: The Weighted Sorted Sample

Monthly Giver?	Fitted Probability	Weight	New %	Monthly Giver?	Fitted Probability	Weight	New %
Yes	0.976	1	0.05	No	0.440	66	43.70
Yes	0.934	1	0.10	No	0.435	66	47.00
Yes	0.897	1	0.15	Yes	0.410	1	47.05
Yes	0.894	1	0.20	No	0.408	66	50.35
No	0.890	66	3.50	No	0.407	66	53.65
Yes	0.877	1	3.55	No	0.386	66	56.95
No	0.865	66	6.85	Yes	0.352	1	57.00
Yes	0.832	1	6.90	No	0.345	66	**60.30**
No	0.805	66	6.95				
Yes	0.778	1	7.00				
Yes	0.749	1	7.05	No	0.339	66	63.60
No	0.744	66	10.35	No	0.327	66	66.90
No	0.710	66	13.65	No	0.324	66	70.20
No	0.650	66	16.95	No	0.288	66	73.50
Yes	0.648	1	17.00	Yes	0.265	1	73.55
Yes	0.646	1	17.05	Yes	0.245	1	73.60
Yes	0.645	1	17.10	No	0.233	66	76.90
No	0.644	66	**20.40**	No	0.205	66	**80.20**
No	0.626	66	23.70	No	0.197	66	83.50
Yes	0.611	1	23.75	No	0.190	66	86.80
No	0.592	66	27.05	No	0.181	66	90.10
No	0.565	66	30.35	No	0.172	66	93.40
Yes	0.543	1	30.40	No	0.156	66	96.70
Yes	0.537	1	30.45	No	0.097	66	**100.00**
No	0.519	66	33.75				
Yes	0.514	1	33.80				
No	0.492	66	31.10				
No	0.471	66	**40.40**				

Table 6.3: Lift Chart Calculations Corrected for Oversampling

Quintile	Incremental Response	Cummulative Captured Response
1	12/408 = 2.94%	12/20 = 60%
2	4/400 = 1%	16/20 = 80%
3	2/398 = 0.5%	18/20 = 90%
4	2/398 = 0.5%	20/20 = 100%
5	0/396 = 0%	20/20 = 100%

That in turn moves the quintile cut points down in the table, and the first bin will now include more of the "yes's" than the first quintile did in the sample. This in turn raises the lift for the earlier part of the lift charts so that the reduction in the incremental response will be less than 40 times; the later parts of the chart will be reduced more than 40 times. The base rate of 1%, however, is exactly 40 times less.

Table 6.3 takes the results directly from Table 6.2 to calculate the incremental response rates and the cumulative captured response. For example, in the best 20% of the database, we can expect to have 12 members from the 408 individuals, for a response rate of 2.94%. Since there are a total of 20 members, we have captured 60% in the best 20% predicted by the model.

Figures 6.4 and 6.5 show the incremental response rate lift chart and the cumulative captured response lift chart corrected for oversampling. These lift charts indicate what we should expect when the predictive model is applied to the complete database.

6.2 Using Lift Charts

One real benefit of lift charts is that they make data mining models directly applicable to managerial decision making. The incremental response rate chart quickly allows a manager to determine how many existing customers should be targeted with a particular sales offer, and the predicted probabilities indicate which specific customers should be targeted. The total cumulative response chart allows the manager to forecast sales, revenues, and profits associated with a particular sales offer. In addition, the total cumulative response chart is a valuable tool for the marketing analyst for selecting a specific predictive model out of a set of potential predictive models. The tutorial for this chapter describes how to use the total cumulative response chart as a tool to select

Figure 6.4: The Weighted Sample Incremental Response Rate Chart

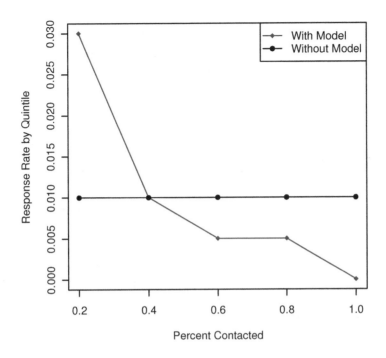

Figure 6.5: The Weighted Sample Total Cummulative Response Rate Chart

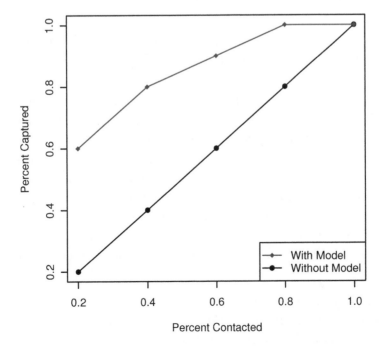

among potential predictive models, while here the focus is on the use of the chart to forecast sales, revenues, and profits.

Direct marketing is the area in which these tools have had their greatest impact. Most direct marketers will run experiments on sales offers that they are considering by sending the offer to a random sample of their customer list to determine which customers to target in the overall database. For example, consider a regional telephone service provider (such as Telus) that is interested in expanding its DSL customer base among its current residential land-line customers who are not current DSL subscribers, a target audience of 1.5 million customers for the firm. The specific appeal for the DSL service includes a three-month reduction in the price of the service and a free DSL modem. Because of the nature of this offer, the regional telephone service provider's marketing managers have decided that the DSL campaign should make the offer in a direct mail piece. Prior to the full launch of the campaign, the company sent the DSL offer direct mail piece to a random sample of 15,000 customers who are in the campaign's target audience of current residential land-line customers who do not subscribe to the DSL service. The cost of each direct mail piece is $2.50, and the expected net present value of the profit contribution from a new DSL subscriber is estimated to be $125. Out of the 15,000 in the sample of target customers who were sent the direct mail DSL offer 150 (or one percent) became DSL subscribers. Consequently, if all 1.5 million target customers were sent the direct mail piece, then (based on the positive response rate among sample customers) the company should expect that one percent of them (or 150,000) would become DSL subscribers.

The expected net present value of $18,750 from the 150 customers (at $125 per customer) in the sample who became DSL subscribers did not cover the $37,500 cost of mailing the DSL offer to the 15,000 customers (at $2.50 per customer), so mailing the DSL offer to the entire target audience appears to be a money losing proposition. However, the $18,750 loss associated with the sample mailing is best viewed as a research cost rather than as a business loss. While the DSL offer should not be mailed to the entire target audience, it seems possible that a more targeted mailing may well be profitable. To undertake the targeting, the sample of 15,000 households is used to develop a predictive model. The sample is broken down into estimation, validation, and holdout sub-samples; and the number of positive responders is oversampled. Figure 6.6 contains the incremental response rate chart for the holdout sample, while Figure 6.7 contains the total cumulative response chart for this sample.

In general, the last customer receiving the direct mail piece should be the one for which the expected contribution from that customer equals the cost of the mailing. What causes the expected contribution to vary across customers is

Figure 6.6: The Incremental Response Rate for the DSL Subscriber Campaign

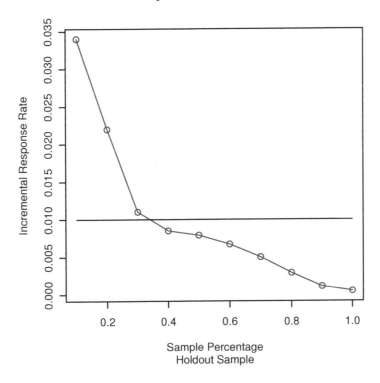

the probability they will respond favorably to the direct mail DSL promotional offer. Let the probability of becoming a DSL subscriber for the "break-even" customer be π, then it must be the case that

$$\$125(\pi) = \$2.50,$$

or, after a little algebra,

$$\pi = \$2.50/\$125 = 2\%.$$

The probability of a positive response and the incremental response rate turn out to be one in the same thing. As a result, using the incremental response rate chart (Figure 6.6), a pencil, and a ruler, one finds that the "best" (based on the predictive model) 22% of the target customers should be contacted. To determine this, draw a horizontal line from 0.02 (or 2%) on the vertical axis of Figure 6.6 to the incremental response rate curve, and then at the point where your horizontal line intersects the incremental response rate curve, draw a vertical line down to the horizontal axis. Your vertical line should intersect the horizontal axis at roughly 0.22 (or 22%). Based on the total

Figure 6.7: The Cummulative Total Response Rate for the DSL Subscriber Campaign

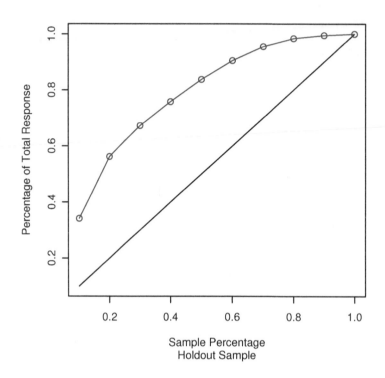

cumulative response chart in Figure 6.7 (along with the ruler and the pencil), the "best" 22% of the target customers are expected to produce roughly 58% of the total new DSL subscribers that would be obtained if all the target customers were contacted. To see this, locate 0.22 along the horizontal axis of Figure 6.7 (where 0.22 corresponds to 22% of the total database), and then draw a vertical line from the horizontal axis at that point to the weighted total cumulative response curve. Next, draw a horizontal line from the point where your vertical line intersects the cumulative response curve out to the vertical axis. You should find that your horizontal line crosses the vertical axis at around 0.58 (or 58%). The 22% of customers to contact translates into $22\%(1.5\text{Mil} - 15{,}000) = 326{,}700$ actual customers that should be sent the direct mail piece (the 15,000 corresponds to the customers in the experimental sample who have already received the direct mail piece), while this mailing is expected to generate $58\%(15{,}000 - 150) = 8{,}613$ new subscribers, where the 15,000 corresponds to the 1% expected response rate multiplied by the 1.5 million target customers, and the 150 corresponds to the new subscribers already generated from the experimental mailing. The cost of mailing the offer to the 326,700 targeted customers (at $2.50 each) is $816,750, while the total

Figure 6.8: The Lift Chart Dialog Box

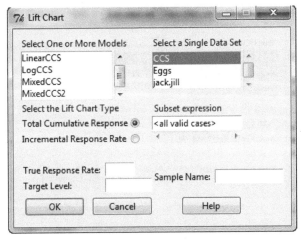

contribution from the expected 8,613 new DSL subscribers (at $125 each) is $1,076,625, resulting in an expected profit from the campaign of $259,875.

6.3 Lift Chart Tutorial

In this tutorial you will learn how to create lift charts using R Commander's lift chart tools, and how to use the validation sample total cumulative response chart to select among possible candidate models.

1.

Double-click on the workspace in which you saved your work from the logistic regression tutorial. This should launch R.

2.

Initially we will be creating a total cumulative response lift chart using the estimation sample. To do this select the pull-down menu option **Assess** → **Graphs** → **Lift charts**..., which will bring up the Lift Chart dialog box shown in Figure 6.8.

3.

In the Lift Chart dialog box **select the four CCS models from the "Select One or More Models" option**. For the **"Select a Single Data Set"**

Figure 6.9: The Completed Lift Chart Dialog

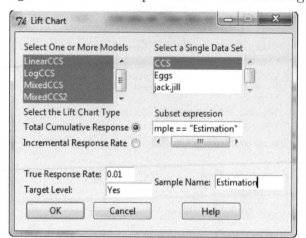

option select **CCS. In the "Subset expression" field enter Sample == "Estimation."** The default Total Cumulative Response for "Select the Lift Chart Type" is correct. The monthly givers in the CCS data set are over-sampled so that they represent 50% of the sample. Recall, though, that they represent only 1% of all CCS donors. Consequently, **enter 0.01 in the "True Response Rate:" field.** The target level (i.e., the desired value of the target variable MonthGive) is "Yes," so **enter Yes (without quotes) in the "Target Level:" field.** The first lift chart we will be making is for the estimation sample. While the final model will be selected based on the *validation* sample, we will first briefly look at a cumulative total response chart for the *estimation* sample. This will give some indication of how the four different models estimated in the logistic regression tutorial perform. Give this chart a title by **entering "Estimation" in the optional field "Sample Name."** This name field does nothing more than add a title, but that is very useful to help you keep track of your charts. Once you have done all of this your Lift Chart dialog box should look like the one shown in Figure 6.9. When it does, press **OK**, and the total cumulative response chart shown in Figure 6.10 should appear in the R graphics window.

4.

An examination of Figure 6.10 indicates that the total cumulative response curve for the MixedCCS model is noticeably below the total cumulative response curve for the other three models over the range of the sample percentage from 10% to 30%, indicating that its performance over this range is inferior to the other models over this range. Similarly, for values of the sample percentage over 50% the LinearCCS model underperforms the other three

Figure 6.10: The Total Cumulative Response Rate Chart for the Estimation Sample

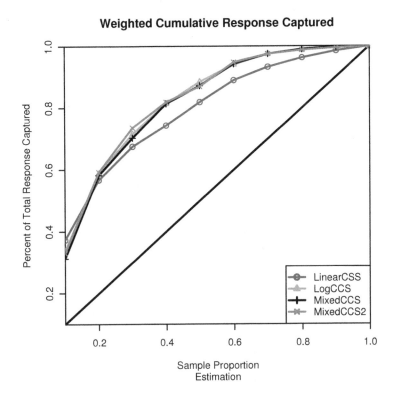

models. The `LogCCS` model and the `MixedCCS2` models perform comparably well over the entire range of the sample.

5.

To select among possible models, we will use the validation data to avoid overfitting. To do this we use the estimated coefficients just calculated based only on the *estimation* data set, and the predictor variables from the *validation* data set to calculate a predicted target probability (of `MonthGive == "Yes"`) for each individual in the validation data set. This predicted probability of the target in the validation data, and the the true observed target value in the validation data, is used to construct the lift chart and *validate* the model. To do this, use the pull-down menu option **Assess → Graphs → Lift charts...** again. **Select the four models and CCS data set**. The only difference between this cumulative response chart and the first one you created is that you need to change the "Subset expression" to Sample == "Validation," enter Validation in the "Sample Name:" field. Once you have done this, and

Figure 6.11: The Total Cummulative Response Rate Chart for the Validation Sample

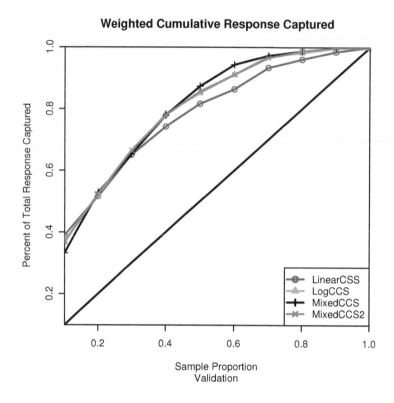

pressed **OK**, you should have a lift chart identical to the one shown in Figure 6.11.

6.

The resulting story for the validation sample is very similar to the one for the estimation sample. The LinearCCS model does best at first (but only slightly better than the LogCCS and MixedCCS2 models), and then underperforms the other three models. The major difference between the two samples is for the MixedCCS model, which now performs worse the other three models for the first 10% of the database, but outperforms the other models from 50% to 70% of the sample. Finally, the performance of the LogCCS and MixedCCS2 models is nearly identical. However, the MixedCCS2 model is preferred to the LogCCS model since it has fewer estimated coefficients (a property known as "parsimony"), and more parsimonious models tend to have better prediction capabilities across multiple new samples of data. Overall, making a decision of which model to select as the "champion" is not easy in this circumstance. Generally, in evaluating models, we place more weight on the ability to predict the "best" prospects (suggesting the selection of the `LinearCCS` model as the champion). However, elsewhere, this model performs quite a bit worse than the other three candidates. Almost the reverse situation holds for the `MixedCCS` model. From the point of view of consistent prediction capabilities, both the `LogCCS` and `MixedCCS2` models are a good choice as the champion model, with the `MixedCCS2` model being the better choice for reasons of parsimony. Consequently, for the purpose of this tutorial, we will select the `MixedCCS2` model as the champion.

7.

Finally, create an incremental response rate chart for the `MixedCCS2` model using the validation sample. To do this, again use the pull-down menu option **Assess → Graphs → Lift charts. . . .** This time **select only MixedCCS2 for the "Select One or More Models" option, select CCS for the "Select a Single Data Set" option, choose incremental response rate as the lift chart typ**e. Again, the remaining fields should already contain the correct information. Once you have completed filling in the Lift Chart dialog box, press **OK** and the incremental response rate chart shown in Figure 6.12 should appear in the R graphics window. Since we have now sorted the

Figure 6.12: The Incremental Response Rate Chart for the Validation Sample

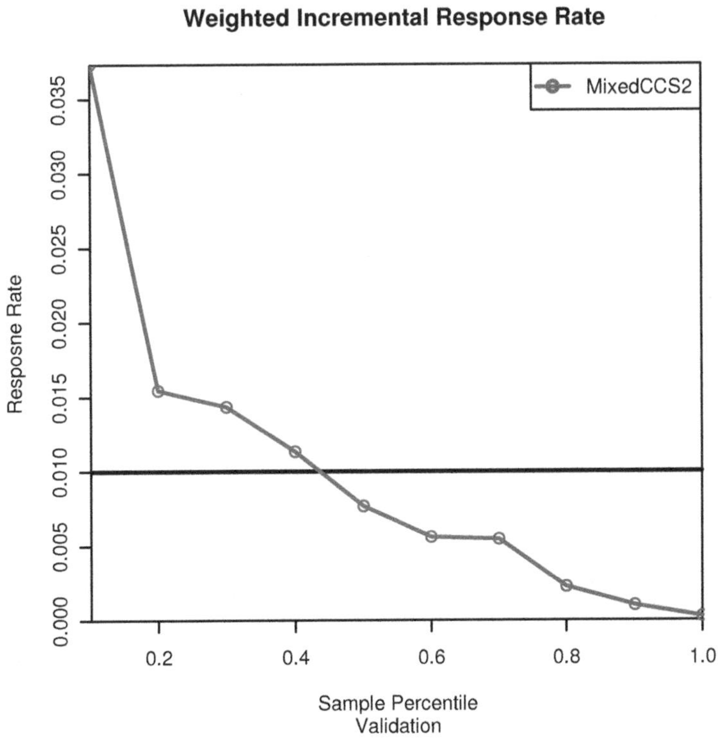

Weighted Incremental Response Rate

donors by the likelihood of joining the Monthly Giving Program, we can see that targeting the most likely donors first will give good incremental response rates (3.7% for the first tenth of donors targeted), and the rates will fall off as we target progressively less likely donors. If we know the costs of contact and the average dollar amount of donations, we can use these incremental response rates to calculate the costs and expected donations associated with contacting more and more donors; then we can choose a profit maximizing level of contact. This is good, but we may be able to do better by using different tools and fine-tuning our procedures.

Chapter 7

Tree Models

One method of predictive modeling, which we call tree models, but are also commonly referred to as decision trees or simply trees, generates a set of if-then rules that allow the use of predictor variables to predict target variables or make a decision. As a simple example of these types of decision rules, consider a rule used by a credit card company deciding whether to issue a Platinum Card to an applicant:

- "If monthly mortgage to income ratio is less than 28% and

- months posted late is less than 1, and

- salary is greater than $45,000,

- then issue a platinum card."

The set of if-then rules for making the decision to issue an applicant a credit card can be graphically portrayed using a tree representation, as shown in Figure 7.1.

Figure 7.1 portrays an actual decision tree. A natural question to ask is how were the if-then rules used in the tree determined? Ideally, they would be derived from historical data for other customers, and would be based on the characteristics of those customers that determined whether or not a particular customer generated a profit or loss for the credit card company. This is precisely what a tree model does. Specifically, these methods infer the set of if-then rules (known as *splits* or *splitting rules*) that separate profitable from unprofitable customers.

In this chapter we describe one algorithm commonly used to determine splitting rules for tree models. Following this, we show you the steps involved in applying this algorithm to a predictive modeling problem using R and R Commander.

Figure 7.1: A Tree Representation of the Decision to Issue a Platinum Credit Card

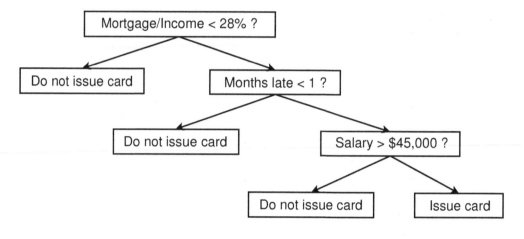

7.1 The Tree Algorithm

To create a tree the algorithm uses an estimation data set with historical information on both predictor and target variables. The target variable is typically binary, which in this case might take the value of "Bad" (to indicate that "this customer has proven to be a costly cardholder") or "Good" (to indicate that "this customer has proven to be a profitable cardholder"). The algorithm searches for decision rules based on the predictor variables (mortgage to income ratio, months late, salary) to progressively divide the cases in the data into smaller groups. The criterion used at each split is that each smaller group has a higher concentration of either "Bad" or "Good" than the preceding larger group. We say that each subset is more homogeneous, or pure, than its parent set. The final groups, called leaves, will have either mostly good or bad cardholders. The decision rules can then be used to sort new applicants who currently do not have a platinum card. The decision rules determine whether the applicant is likely to be a good or bad customer based on the leaf to which they are assigned, and provide the corresponding decision as to whether or not to issue a card.

It may be helpful to compare tree models with logistic regression:

1. Both logistic regression and tree models require historical data containing both predictor and target variables.

2. Both regression and tree algorithms use the historical data to create (i.e.,

calibrate) a model that approximates the relation between the predictor and target variables.

3. The regression model is an algebraic formula, while the tree model is a sequence of if-then rules.

4. Either model, once generated, can then use predictor variables from a new case and "predict" the likely value of the target variable.

5. The logistic regression model calculates the probability of a target variable taking the value "Good" for a particular case (e.g., customer). This probability can then be used to split the data into two groups using only the predictor variables. For example, we can separate the data into cases that are more likely to have a target value of "Good" from those that are more likely to have a target value of "Bad" by splitting at the probability of 0.5. The tree model can directly assign a new case (e.g., a customer) to a group (leaf) that is more likely to have either a "Bad" value or a "Good" value for the target variable. The probability of a new case actually being a "Good" (or "Bad") is given by the proportion of "Good" (or "Bad") customers in the assigned group (leaf) in the estimation data set.

Tree models have both advantages and disadvantages over other types of predictive models. Interpretation of a calibrated tree model is very simple and intuitive, which makes it particularly useful for presentations to non-technical audiences. It is also more flexible than regression models in its assumptions about model structure, easily handles large numbers of variables, and is less affected by missing values. These features make it a good tool for initial exploration of a database. However, in our experience, a regression model will pick up weaker (but still meaningful) effects; and with enough data, the right variable transformations, and careful management of missing values, will fit and predict as well (and typically) better than a tree model.

Tree model algorithms vary in a number of characteristics, including the type of target variable they handle and the branching complexity. We will use the *rpart* (which stands for *recursive partitioning*) algorithm (Venables and Ripley, 2002), and confine ourselves to binary target variables and binary branches. The *rpart* model is also known as the Classification and Regression Tree (CART or C&RT) model (Breiman et al., 1984).

7.1.1 Calibrating the Tree on an Estimation Sample

At each stage of tree "growth," the rpart algorithm searches all possible split points of all possible predictor variables to find the one that splits the cases

Figure 7.2: A Three-Node Tree of Potential Bicycle Purchases

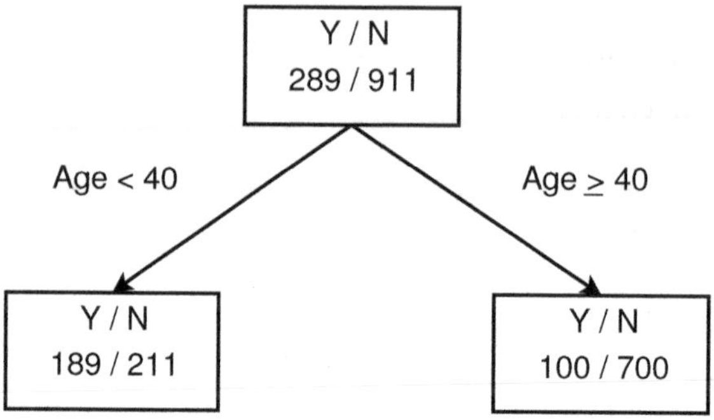

into the most homogeneous (or "pure") subgroups. This requires a measure of purity so that all possible splits can be compared. The measure we will use in rpart is what is known as the Gini index. The following example illustrates the calibration process.

Suppose a bicycle shop has collected information on adult customers entering the store. This information includes the customer's age, the number of children he or she has, and whether or not he or she purchased a bicycle. The owner of the shop wishes to quantify the relation between a bicycle purchase — the target — and age and children — the predictors. The data set has 289 buyers and 911 non-buyers, or 24% buyers out of the total. This information is contained in the variable "Buy" which takes the values "Yes" or "No." Using the decision tree strategy of searching for binary splits, we would start generating trial splits. For example, a trial split at customer age 40 produces 400 customers who are under 40, and 800 who are over 40, with the following proportions of buyers and non-buyers:

Buy	Age < 40	Age ≥ 40
Yes/No:	189/211	100/700
Total in branch:	400	800
	(47% Yes)	(12.5% Yes)

We can write this as a tree with three *nodes* (Figure 7.2). Leaves are the final nodes in a calibrated model.

This looks like an improvement—the concentration of buyers in the under-40 group, 47%, has doubled over the original concentration across all customers, while the 40-and-over group has become much more "pure" than the original group (at 12.5% buyers compared to 24% for the group overall). In order to automate the decision as to the best split, we need a single number that we can use to compare the improvement across all possible splits, and use that number to pick the best split. As a guide for finding such a measure, a 50–50 split is the most impure a node can be, and a 0–100 (or 100–0) split is least impure. The Gini index provides such a measure of the change in impurity resulting from a split of the data. The Gini index is calculated as follows for the bicycle shop example:

Define P_L and P_R as the proportions of the original node in the left and right nodes, resulting in

$$P_L = 400 / 1200 = 0.333$$

$$P_R = 800/1200 = 0.666.$$

Define the proportion of one type (Yes) within the left and right subnodes as p_L and p_R, which gives

$$p_L = 189/400 = 0.4725.$$

$$p_R = 100/800 = 0.125$$

Define the Gini index as

$$P_L(p_L)(1 - p_L) + P_R(p_R)(1 - p_R)$$

$$= 0.333 \times 0.4725 \times 0.5275 + 0.666 \times 0.125 \times 0.875$$

$$= 0.154.$$

We can calculate this single number for all possible splits on all possible predictor variables. To build some intuition for this index, note that when p_L or p_R are 0.5 (the most impure, 50–50, split) the terms $(p_L)(1 - p_L)$ and $(p_R)(1 - p_R)$ are largest, having a value of 0.5 times 0.5 = 0.25. When p_L or p_R is one or zero, the most pure cases, then these terms are smallest, having a value of zero. These terms are then combined into a weighted average, with P_L and P_R as the weights. The most pure subgroups have the smallest Gini index.

For the bicycle shop, there are many possible splits on the age variable, and one possible split on the kids variable. Table 7.1 provides the Gini index for six possible splits, five for the age variable and the one possible split for the kids variable.

The first split in the tree would be on age, at a value 40 (since this has the lowest associated Gini index). All of those customers under 40 go into one group, and all of those 40 or older into the other group. In this instance, the split is driven by the combination of the size of the over-40 group (two-thirds of potential customers entering the store) and the high degree of purity (12.5%) associated with this group after splitting. The under-40 group is actually less pure, but it is half the size of the 40-and-over group. The same procedure would then be followed to split each of these subgroups into two further groups, most likely splitting the under-40 group given its high impurity level at the next split. The process continues until a stopping rule is met at each node.

7.1.2   Stopping Rules and Controlling Overfitting

A node is not split again, and becomes a leaf, when a stopping rule is met. Some stopping rules are:

- All cases in a node have identical values for predictors.

- The depth of the tree has reached its pre-specified maximum value.

- The size of the node is less than a pre-specified minimum node size.

- The split at a node results in producing a child node whose size is less than a pre-specified minimum node size.

- The node becomes pure, i.e., all cases have the same value of the target variable.

- The maximum decrease in impurity is less than a pre-specified value.

- The tree exceeds a pre-specified level of complexity as measured by a complexity parameter, cp.

Table 7.1: Possible Splits for the Bicycle Shop Customer Data

Possible Split	Gini Index Value
Kids="no," Kids="yes"	0.213
Age<30, Age$\geq$30	0.190
Age<40, Age$\geq$40	**0.154**
Age<50, Age$\geq$50	0.168
Age<60, Age$\geq$60	0.208
Age<70, Age$\geq$70	0.219

In our R Commander implementation of *rpart*, the stopping rule under the analyst's control is the complexity parameter. Other rules are not used, or left at their default values.

In regression modeling we can get an increasingly better fit of the data by incorporating more predictor variables, and by including more complex and flexible transformations of those variables. With trees models we can progress with increasingly finer splits to grow larger and more complex trees. In both cases we will eventually be fitting noise unique to that particular estimation data set. As a result, we need criteria that prevent us from fitting noise. The approach used by *rpart* is to grow a large (over-fitted) tree and, at each stage of growth and increasing complexity, to automatically generate holdout samples as subsets of the estimation sample. These automatically generated holdout samples are used to validate each of the sequence of tree models. That is, the model is calibrated on the data not including the holdout data. This model is then used to classify holdout data. The validation error is the misclassification rates in the holdout data. This *cross-validation* procedure generates a sequence of measures of fit, the cross-validation error, for increasingly more complex trees. The analyst can then use the cross-validation error to *prune* those branches of the tree that appear to be fitting noise.

As indicated above, the parameter that the analyst actually sets in rpart to determine which branches to keep and which to prune is called the complexity parameter, designated cp. In one way this is an unfortunate label, since the complexity (or depth) of the tree actually increases as the cp decreases — it might more logically have been called a simplicity parameter. The algorithm outputs a table and a plot that gives the number of splits, the cross-validation error, and the complexity parameter at different levels of the tree. As with validation error generally, it is typical to observe the error decrease with increasing complexity, stabilize, and then start to increase again as the model becomes overfitted. The table or plot can then be used to choose the value of cp that gives the desired degree of fit. The algorithm is then rerun with this cp value entered to produce the final model.

Figure 7.3 shows a cross-validation error plot. The vertical axis gives the relative cross-validation error, where "relative" means relative to the root node (a trunk with no branches). This results in a natural fit measure for a particular tree size as a percentage of the error that would occur if no splits were used. This normalization is accomplished by dividing the error for each tree by the error of the root node. The horizontal axis is labeled by both the complexity parameter (along the bottom), and the number of nodes (along the top). As the number of nodes, and hence complexity, increases, the cross-validation error decreases, levels out, and then increases as the model becomes overfitted. From this plot the appropriate value of cp can be chosen to generate the final

Figure 7.3: A Relative Cross-Validation Error Plot

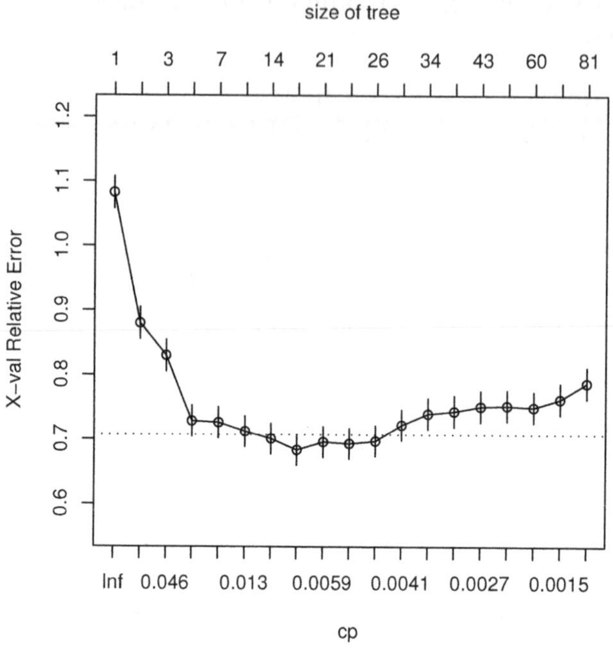

model. One choice is at the minimum error, corresponding to 17 nodes and a cp value of about 0.008. However, there are several other values nearby that are very similar, and each value is an estimate, so that the notion of being statistically equivalent applies. In an analogy to a confidence interval, the algorithm also calculates the standard deviation of the relative error at the minimum value, and plots a horizontal line that is one standard deviation away from the estimated value. Any point below that line is said to be statistically equivalent to the minimum point. If we value parsimony, as most modelers do, we would choose the simplest tree that has a cross-validation error statistically equivalent to the minimum error, which in this case corresponds to the tree with 14 nodes.

7.2   Trees Models Tutorial

This tutorial will demonstrate the use of the *rpart* (*recursive partitioning*) algorithm to generate a tree model for the Canadian Charitable Society data, and will introduce you to some of strengths and weaknesses of tree models in comparison to logistic regression models.

Figure 7.4: The rpart Tree Dialog

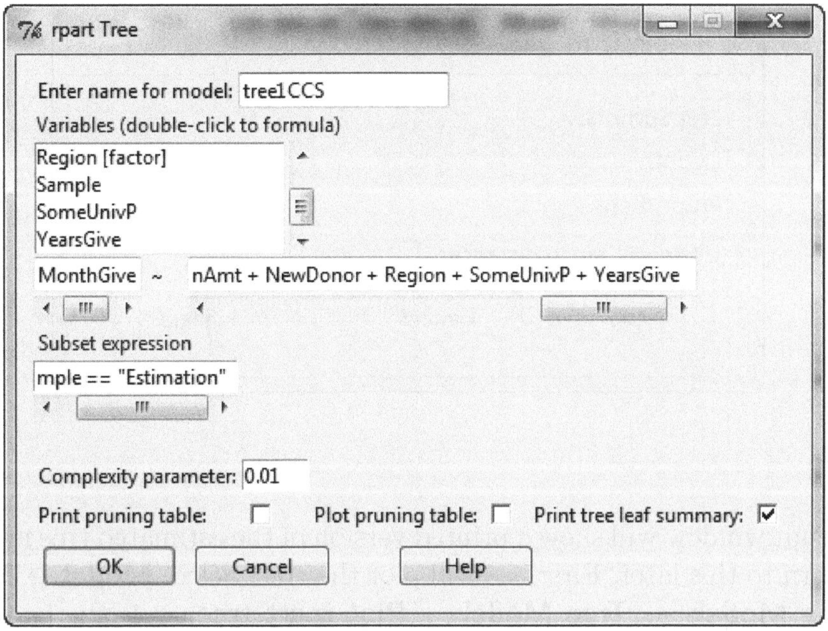

1.

Start R and R Commander, load the BCA package and the CSS data set, and **create 50/50 estimation and validation samples** as you did in the logistic regression tutorial. **Click on Model → Machine Learning Models → Tree Model → Train rpart tree...** to bring up the dialog box shown in Figure 7.4. **Enter `tree1CCS`** as the name for the model. The target and input variables are selected in the same way as in the regression model: **Choose `MonthGive`** (the factor) **as the target, and all of the remaining original** variables (not the new computed variables, if you have retained them from the last tutorial, nor Sample) as predictors. As usual, we will wish to validate our estimated model later, so **set the sample to Estimation**. The *complexity parameter* determines a cutoff level for how weak of predictors you wish to include in the tree, and thus how large a tree you wish to grow. To include weaker effects in the final model, the parameter is set closer to zero. As usual with data mining, including progressively weaker effects in the model means that at some point weak effects will be unique to this particular data set (noise), and will be useless when applied to a new data set. A reasonable value requires some effort (and a bit of judgment) to determine, which we will explore later. For now, **leave the complexity parameter at the default value of 0.01, uncheck the Print pruning table: and Plot pruning table: options. Click OK.**

Figure 7.5: The rpart Tree Plot Dialog

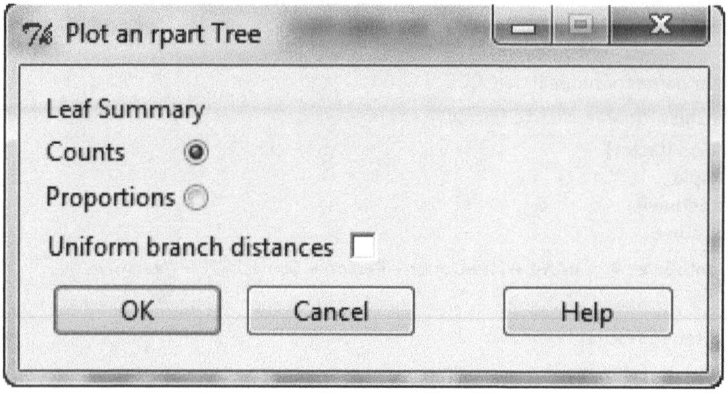

2.

The output window will show a printed version of the estimated tree model. We will return to this later. First, we will plot the tree. Select **Model → Machine Learning Models → Tree Model → Plot rpart tree...** from the menu, to bring up the dialog in Figure 7.5. **Uncheck Uniform branch distances** and click **OK**, and the plot in Figure 7.6 should appear.

3.

The plot of the tree in Figure 7.6 shows the approximate split values that the model finds. Precise values can be seen in the text output. The best first split of the data to concentrate the target `MonthGive` is at the `AveDonAmt` of about $12 (the text output shows it more preciesly as $11.83). The convention in the plot is that those individuals where the statement in blue is *true* go to the left branch. In this case, those with **AveDonAmt < 12** go to the left branch, and individuals with an average donation amount greater than $12 go to the right branch. The vertical length of the branches roughly indicates the relative strengths of the relationships between the splitting variables and the target variable. In this case, the first split is by far the strongest.

The final nodes are called *leaves*. The right branch splits again on AveDonAmt, with those indivduals donating more than about $24 moving to a final leaf. This leaf is labeled "Yes" which indicates that the majority of individuals joined the monthly giving program. The specific counts, 259/365, are then given: 259 individuals joined the program out of 365 who had both preceding conditions satisfied. This makes good intuitive sense: if the average donation amount is larger, donors are more likely to join the monthly giving program. The far left branch is split again on Donations per year, with its left branch taking those individuals with the true condition, specifically with donations

Figure 7.6: The CCS Tree Diagram Where Branch Length Indicates Importance

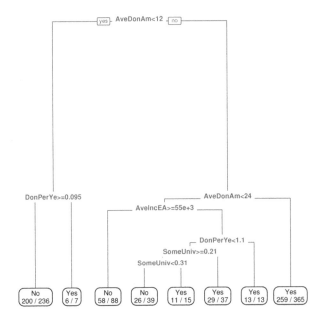

per year greater than 0.095. The far left-hand leaf labeled "No" indicates that the majority of individuals in that leaf did not join the monthly giving program. The numbers provide the counts: 200 out of 236 in this group did not join the program. These 236 individuals satisfy the previous two conditions, that is, they give more than 0.095 times per year (i.e., about once every 10 years) *and* had an average donation amount under 12.

Note that very small and very large numbers are given in scientific notation, e.g., "e+3" means "times 10 raised to the power of three," or times 1000.

Now we can use the tree model to predict the likelihood of a new individual donor joining the monthly giving program. All we need is the value of the predictor variables for that individual. Then we can follow the path through the tree, with the prediction for that individual being the label on the leaf that the individual ends.

Exercise: Using the calibrated tree model, predict whether or not the following new donors are likely to join the monthly giving program:

Donor #1: Ave DonAmt = $19, AveIncEA = $85,000

Donor #2: AveDonAmt = $9, DonPerYear = 0.04

Donor #3: AveDonAmt = $22, AveIncEA = $52,000, DonPerYear =0.04, SomeUniv = 0.25

Figure 7.7: The CCS Tree Diagram with Uniform Branch Sizes

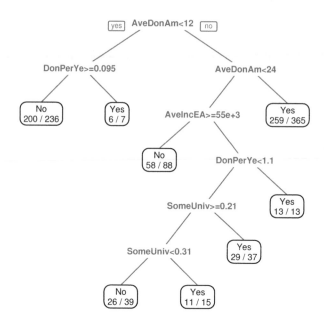

4.

Plot display options: Drawing the tree with the branch length indicating relative strength of the relationship is initially useful, but it usually crowds the remaining branches, making them difficult to read. The tree can be drawn with uniform spacing to make reading the labels easier. To do this, **bring up the Plot rpart tree training dialog box** again but this **time leave "Uniform branch distances"** checked. **Press OK**. Figure 7.7 shows the redrawn tree.

Recall that Windows users can save plots by **clicking on the History menu** at the top of the graphics window, and **selecting "Recording."** If you do not record the plots, each new plot will overwrite the previous one, losing it. You can also redo the plots with each leaf showing the proportions of the two values of the target variable, by **selecting "Proportions"** in the plot dialog. Try this now. As long as the previous plot was saved, you can **switch between the two plots using the Page Up and Page Down keys**.

Exercise: In one sentence, explain the meaning of the two proportions given in a node.

5.

Tree Table: In the R Commander output window, **scroll back to the output from the tree model**. The text output gives a list of the variables used

Figure 7.8: The Printed Tree

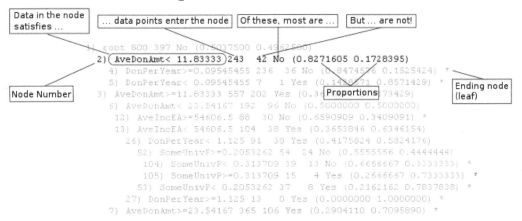

for the given complexity parameter, a table version of the tree, and a "pruning table." The table version of the tree is shown in Figure 7.8, along with explanatory captions. The root node does not have split information, so the captions are directed to the second node. To read the table, use the captions, and fill in the "..." with the entry in the table. The node numbering starts with "1)" for the root, which in the table includes all those individuals entering the first split. To help follow the branching system, the convention is that the left branch in the tree plot has a node number that is double the preceding node, and the right branch is double plus 1. Hence the branches from node 3 are nodes 6 and 7. This makes the node number increase rapidly, but it does aid navigation of the tree table and comparison with the graphic tree. Node 2) in the table thus corresponds to the first node in the left branch of the split. Notice that the condition given in the table is for individuals ENTERING node 2), which in the tree is the label on the preceding node. Thus, node 3) has the opposite condition from node 2). The main piece of additional information beyond what is in Figure 7.7 is the proportions of the binary variable in the node. Compare the information in nodes 2), 3), 4), and 5) with Figure 7.7.

Exercise: The tree *plot* only gives the count information for leaves. The Tree *table* gives count information for all intermediate nodes as well. Using the tree table, determine what the count information would be on the node in the tree plot that is labled "AveIncEA>=55e+3" and for the node labled "SomeUniv>=0.21"

In the next two steps we will explore tree models' characteristics of structural flexibility and the effect of correlated predictors. Following that we will learn how to assess and control overfitting in tree models, before going on to model comparisons using lift charts.

6.

Structural Flexibility: In logistic regression, a natural logarithm transformation of the `AveDonAmt` variable improved the fit because there is a convex "decreasing returns" relationship, and because the untransformed model makes strong assumptions about the structural relation between target and predictor variables that does not include the convex relationship. Tree models make weaker assumptions, one consequence of which is that trees handle the nonlinear relation between `AveDonAmt` and `MonthGive` *without the need for a transformation.* To demonstrate that the log transformation is not necessary, we will run the model with the transformed variable. **Create a new variable, Log.AveDonAmt, the log of AveDonAmt** as in the logistic regression tutorial. Use **Data → Manipulate Variables → Compute New Variable. Rerun the tree with the same parameters as before** except using `Log.AveDonAmt` but **not including AveDonAmt**. Since the dialog box has the previous model parameters, you can quickly make the change by placing the cursor in the input variable box, using the right arrow on the keyboard to move to `AveDonAmt`, and just typing the prefix "`Log.`" to it. **Name** the model **trlogADA,** and **click OK. Plot** the new tree using the default plotting parameters. The result is shown in Figure 7.9. For Windows users only, you can select **History** at the top of the Graphics Window menu, and use **Previous** and **Next** to go back and forth between the two trees (or use the Page up /Page down keys). Note that all that changes in Figure 7.9 is the value of the split. It is the log of the value of the split in Figure 7.7. The same individuals end up in the same leaves.

Generally, any *monotonic* transformation (one which does not change the ordering of the values) of a variable will not change the resulting tree model. While this flexibility in handling nonlinear relations between predictor and target is a strength of tree models, particularly for exploring data, it is difficult to interpret nonlinearities beyond the first or second split, so the various plotting methods and analyses discussed in previous tutorials remain useful for exploring data.

Delete the Log transformed variable (using the Data→ Manipulate Variables → Delete Variables from the Data Set menu), as we will no longer need it and do not want to accidentally include it later on.

7.

Correlated predictor variables

The dominant feature of Figure 7.7 is that the first split on `AveDonAmt` gives a much greater improvement in fit than any subsequent split. From the logistic regression tutorial, we know that this variable and `LastDonAmt` are

Figure 7.9: The Log Transformed Average Donation Amount Tree

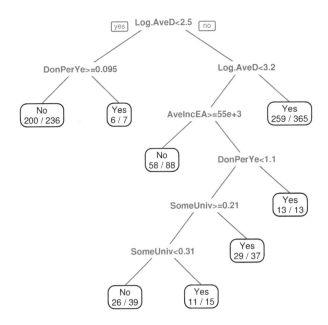

correlated with each other. That means that they would separate donors very much the same way, so that once `AveDonAmt` is used, there is nothing left for `LastDonAmt` to do, and it will not be used. To highlight this, we will rerun the model without `AveDonAmt`. As before, **bring up the tree model training dialog**, which should contain the last model, and **delete** `Log.AveDonAmt` from the command, so that now there is no average donation amount information used as a predictor. **Change the name of the model to TreeNoADA. Run it and then plot it,** using the default values. Figure 7.10 shows the resulting tree. `LastDonAmt` is now able to have a large impact because `AveDonAmt`, which has nearly the same relation with `MonthGive`, is no longer available. What this generally means is that with highly correlated predictor variables one or the other will appear as effective predictors in the tree model, but not both.

In the logistic regression tutorial we found a similar effect in the sequence of logistic regression models. In some models one of these two variables did not even appear as significant, even though they are both individually quite strongly related to the dependent variable. Even in the final model, where both appear, neither is indicated as extremely strongly significant (in fact, `AveDonAmt` appeared to be insignificant). This is a general consequence of correlated predictor variables in the same model: they will reduce the apparent impact of each other on the target. In regression models this happens because the estimate of the variables' variances is increased (a phenomenon known

Figure 7.10: Last Donation Amount Tree

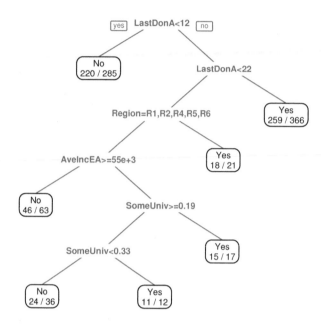

as "variance inflation"), therefore lowering their significance (indicated by increasing their *p*-values). In tree models, whichever variable is used first (e.g., `AveDonAmt`) takes up much of any effect of the second variable, so it is not used. This mainly causes difficulty in *interpreting* models, and can complicate model selection. The good news is that including variables that are correlated with each other will not diminish the *predictive* usefulness of either tree or regression models, as long as the estimated model is applied to new predictor data that has a similar origin (and hence structure) as the original data on which the model is estimated.

8.

Overfitting

As indicated earlier, letting a tree grow too large will result in overfitting the estimation sample. To avoid this, we need to have a stopping rule that prevents us from fitting noise. For trees, the approach used by R is to grow a large (overfit) tree, and then prune it back using a *cross-validation* measure. The CCS data, because of the two strong predictors, `AveDonAmt` and `LastDonAmt`, does not lend itself to demonstrating the logic of the procedure (since the method selects a model with only a single split on one of the two variables), so we will run a tree model without these two strong predictors to demonstrate the use of cross validation and pruning to control overfitting. This time we want to print the cross-validation or "pruning" table and associated plot.

Figure 7.11: The CCS Pruning Table

```
Output Window

Root node error: 397/800 = 0.49625

n= 800

         CP nsplit rel error  xerror     xstd
1  0.2342569      0   1.00000 1.06045 0.035573
2  0.0680101      1   0.76574 0.76574 0.034581
3  0.0377834      2   0.69773 0.70025 0.033925
4  0.0125945      3   0.65995 0.68262 0.033719
5  0.0115869      8   0.59446 0.73804 0.034324
6  0.0100756     13   0.53652 0.76826 0.034603
7  0.0088161     14   0.52645 0.77582 0.034667
8  0.0075567     16   0.50882 0.77834 0.034688
9  0.0062972     21   0.47103 0.79093 0.034789
10 0.0050378     31   0.40302 0.79345 0.034809
11 0.0025189     32   0.39798 0.78338 0.034730
12 0.0012594     36   0.38791 0.79093 0.034789
13 0.0010000     38   0.38539 0.79597 0.034828
```

Bring up the rpart tree training dialog box and generate the appropriate commands (rename the model (noAorLDA); use all the *original* – not log transformed – variables except AveDonAmt and LastDonAmt; set cp = 0.001; all boxes checked; and OK). Look for the pruning table in the output window — you may have to scroll up to see it. The pruning table will be similar to Figure 7.2. The last two columns, xerror and xstd, will be slightly different, and will vary somewhat with every run. The table shows how many splits would have been allowed for various values of the complexity parameter, from no splits, down to the level of cp you set in the dialog box (in this case 0.001, and 38 splits). As mentioned earlier, "complexity parameter" is a rather unfortunate name, as the complexity of the tree *increases* as cp *decreases*.

At each split the improvement in the purity of the tree is given as a fraction of the root impurity, under rel_error. The relative error of the root node (the nsplit= 0 case, in the first line) is set to one. Each subsequent split is chosen as the one that most improves the purity of the tree, and its error continually decreases. This is analogous to the typical increase in R^2 in a multiple linear regression model, or the McFadden R^2 in logistic regression, as more variables are added. Fit always increases, error decreases, and overfitting will occur at some point. To control overfitting, the tree routine also automatically and repeatedly splits the input data set into its own estimation and validation samples, re-estimates the model on the new (smaller) estimation sample, and then uses the model to predict the target in the validation sample. The error between the known and predicted target in these internal validation samples

Figure 7.12: The CCS Pruning Plot

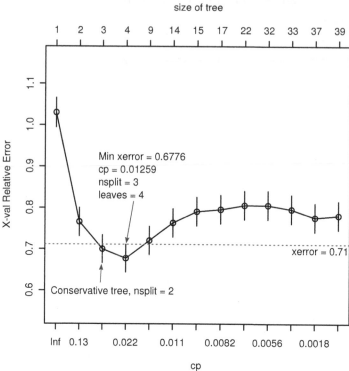

is averaged over several iterations, and is called the *cross-validation error*, and labeled *xerror*. The table shows the common pattern of error measures that are adjusted to compensate for overfitting: xerror decreases, levels out, and then increases, with the minimum indicating approximately where overfitting starts. This is analogous to the behavior of adjusted R^2.

The table has a minimum xerror value of 0.67758 (your may have a slightly different value) at nsplit = 3 and cp = 0.0125945. Not all splits are reported; intermediate splits that could not possibly be candidates for the optimal tree are left out.

In the graphics window, R has also plotted xerror as a function of the size of the tree in terms of the number of leaves (the number of leaves is equal to the number of splits plus one, and is given along the top of the plot as "size of tree"), shown in Figure 7.12. Note that the cp scale at the bottom may not be accurate, and should not be used (free open-source software is sometimes not polished); rely on the values in the pruning table. Recall that the precise value of xerror depends on how the algorithm divides the data into estimation and validation samples during cross-validation. Since this is done randomly, each run will give slightly different results, and your plot may be slightly different.

9.

Pruning the tree: Since each xerror value is an estimate, it comes with a standard deviation, xstd, shown in the last column of the table. The argument for the "best conservative tree" is that any xerror values within one standard deviation of the minimum xerror value are statistically indistinguishable. From the pruning table we see that the minimum xerror is (rounded) 0.678, and has xstd of 0.034. Thus any splits with xerror less than 0.678 + 0.034, about 0.71, are equally good on the basis of cross-validation. This level is indicated by the dotted line. As a result, we consider the 3 and 4 leaf (2 and 3 splits) trees as statistically equivalent. To choose between those two trees, we could add the criterion that we want the most parsimonious (simplest) tree, which will be easiest to interpret. This is the tree with 3 leaves (2 splits). As stated, this is conservative from an overfitting perspective, and we might be willing to give up a bit of easy interpretation for a slightly better prediction, and select a slightly more complex tree without overfitting. Since our original validation sample has not yet been used, we can ultimately use lift charts to compare these two on our validation sample.

Use the pruning table to identify the cp value associated with the desired tree. The most conservative tree, at nsplit = 2, is at cp = .0377834. To generate the conservative tree, rerun the model with any cp cutoff between that cp value and the next larger value (0.0680101) in the pruning table — let's choose 0.04. The next two more complex trees occur at cp values of 0.0125945 and 0.0115869. Setting cp values of 0.013 and 0.012 will give us those trees. **Rerun the model, changing only the complexity parameter to 0.04, and the model name to `noAorLDA04`. Plot and inspect the Rpart tree. Repeat this for complexity parameter values of 0.013 and 0.012,** renaming the models accordingly, and inspecting the plots. Note how the tree grows as cp decreases. Then select **Assess → Graphs → Lift Charts** from the menu. **Select the four noAorLDA tree models and the CCS data set** to see the improvement in the estimation sample. Enter **0.01** for the true response rate, **Yes** for the target, set the Subset expression to **`Sample == "Estimation"`**, and **Estimation** for the name of the sample. The result should appear as in Figure 7.13.

As expected, the more branches in the tree (the smaller its complexity parameter), the better the fit to the *estimation* sample. Next, plot the lift charts on the *validation* sample. **To do this, bring up the lift chart menu again, select CSS as the data, set the "Subset expression" to `Sample == "Validation,"` and label the plot Validation.** Set the remaining parameters as before. Figure 7.14 shows the lift chart. Unlike the last figure, this figure indicates that no tree model dominates the other models throughout the range of the data. In this particular case, the most conservative generally

Figure 7.13: CCS Estimation Weighted Cumulative Response

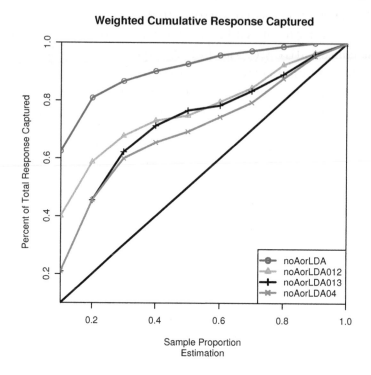

fairs the worst, and the advantage of simplicity in interpretation is not great. Every case will be different, but the large differences between models here and in Figure 7.13 are common, and due to overfitting the noise within the estimation sample by the more complex trees.

10.

Finding a good tree model: Since we are now most interested in good prediction, rather than clear demonstration of the method, we will bring the two strong predictor variables back into the model. **Include AveDonAmt and LstDonAmt with all of the other original variables (not log transformed) and set cp = 0.001. Name the model Full001 and run it.** The pruning chart in Figure 7.15 suggests that anything beyond three splits is overfitting.

Set the complexity parameter to 0.02 (a value between the 0.012 and 0.035, see Figure 7.15) **and rerun, naming the new model Full02.** This will give the minimum xerror model with three splits. **Plot the tree.** This time, we will explore whether going slightly more complex than indicated by the cross-validation plot might perform better if we make our judgment on our own validation sample instead, using lift charts. To get the next most

Figure 7.14: CCS Validation Weighted Cumulative Response

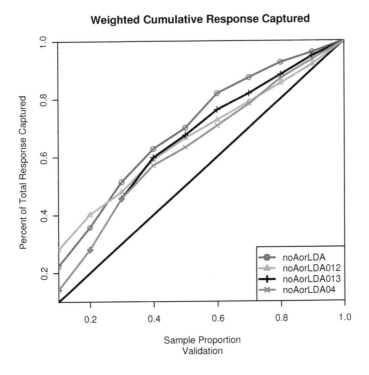

Figure 7.15: The Minimum Cross-Validation Error in the Pruning Table

```
Output Window

Root node error: 397/800 = 0.49625

n= 800

          CP nsplit rel error  xerror     xstd
1  0.3853904      0  1.00000  1.07557 0.035541
2  0.0352645      1  0.61461  0.61965 0.032877
3  0.0125945      3  0.54408  0.57683 0.032203
4  0.0109152      4  0.53149  0.60202 0.032610
5  0.0067170      7  0.49874  0.61713 0.032839
6  0.0062972     12  0.46348  0.64484 0.033234
7  0.0050378     23  0.36272  0.64736 0.033268
8  0.0041982     25  0.35264  0.67506 0.033627
9  0.0025189     28  0.34005  0.69270 0.033838
10 0.0010000     31  0.33249  0.70025 0.033925
```

complex model with four splits, **set the complexity parameter to 0.011, and name the model Full011**. Plot the tree and note where the new branch is.

Now we will compare the lift charts of the various models on the validation set. Select **Assess → Graphs → Lift Charts** and in the dialog box **select the Full02, Full011, Full001**, and just for comparison, **NoAorLDA** from our earlier exercise. Use **the validation sample and set the other parameters as before**. The resulting lift chart (Figure 7.16) shows that the Full011, the 4 split tree, does much better than the overfitting very complex tree, Full001, and slightly better than the simpler three split tree, Full02. The lift chart provides a more detailed validation measure than the cross-validation. Interestingly, at the very beginning of the lift chart, the model without the two strong variables predicts slightly better than any of the others. For now, we will take **Full011** as our "champion" tree model.

Figure 7.16: Weight Cumulative Response Comparison of the CCS Tree Models

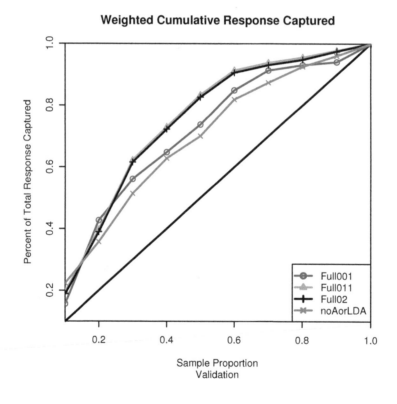

11.

Save your workspace and exit R and R Commander.

Chapter 8

Neural Network Models

Neural networks (Venables and Ripley, 2002) are a class of predictive models that complement the previous models studied. The main advantage of neural networks is their structural flexibility. They can capture an unlimited variety of functional relations between predictor and target variables simply by including more nodes, or neurons, in the model. Their main disadvantage is that a calibrated model, while useful for prediction, is difficult to interpret. The estimated coefficients in regression models and split points in decision trees provide useful information to the analyst even before any attempt at prediction is made. The internal structure of a calibrated neural network model, however, is very complex, and thus not amenable to interpretation. We will first briefly describe the biological inspiration for neural network models, which were initially developed in the field of artificial intelligence (AI), and then their interpretation as a highly flexible statistical model.

8.1 The Biological Inspiration for Artificial Neural Networks

Let's consider two variations on the general problem of aiming a projectile toward a target.

Ballistic Trajectory Problem #1: Field Artillery

An artillery gunner must set the angle of the gun barrel so that the ballistic trajectory of the shell ends on the target to be destroyed. To do this, she enters the distance to the target into a formula that gives the correct barrel angle. The formula is based on classical physics and known parameters such as the muzzle velocity of the shell V_o and the force of gravity g. Figure 8.1 shows the formula and allows anyone to do this common calculation. More sophisticated variations would include air resistance.

Ballistic Trajectory Problem #2: The Jumpshot

A basketball player launches himself into the air, clear of an opposing player, and while airborne sends a ball on a ballistic trajectory toward a hoop 5 meters away. The balls slices cleanly through the hoop, as in Figure 8.2.

Figure 8.1: The Artillery Launch Angle Calculation

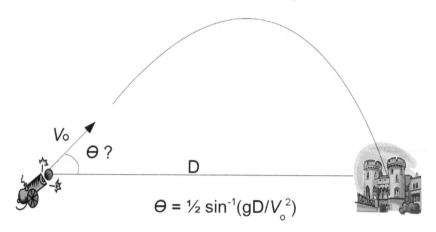

$$\Theta = \tfrac{1}{2}\sin^{-1}(gD/V_o^2)$$

Figure 8.2: The Launch Angle "Calculation" in Basketball

The jump shot calculation is a vastly more complex problem than the artillery calculation. The equations of motion might, in principle, be written down, but there's not much point in the exercise. You can bet the player is not going to measure all of the parameters involved, let alone plug them into a formula, and then execute the necessary motions. His brain, however, must take in that information, process it to provide a predicted output, and give the necessary complex directions via his nerves to his muscles. This all occurs rapidly, automatically, and with a remarkable degree of predictive accuracy. When put in the context of the field artillery problem, which looks like it has some complicated equations, this is an absolutely amazing feat. Yet it has nothing to do with formulae from physics. Rather, it results from learning, by trial and error, over a period of time.

But just how does the brain, which is just a mass of soft, wet nerve cells in a soup of neurotransmitters, do this? That, of course, is the domain of neuroscience, the cognitive sciences, and computer science. Here, we will briefly explain a major insight on the functioning of neurons that led to intensive research in Artificial Intelligence (AI) and ultimately to the Artificial Neural Network (or ANN) we use as a predictive model today.[1]

Neurons in the brain are linked together in a network. Figure 8.1 shows an abstraction of the connections between two neurons. Signals pass from the sending axon across the terminals to the dendrites of the receiving neuron. The neuron will "fire," that is, send a signal through its axon at the right, if it receives enough input from its dendrites. A simple model is that the receiving cell adds up all of the inputs it receives at its terminals from all of the other neurons it is connected to, and fires if that total exceeds a threshold level. That by itself won't do very much. In particular, it won't allow learning to occur. Learning means that behavior changes, and that means that the way the signals are transmitted through the network must change over time.

The additional piece that is necessary to make this work is that *the strength of the connections between the neurons is adjusted when the neurons fire.* More firing strengthens the connections, effectively reducing the threshold level required for the next neuron to fire. Less firing weakens the connections, effectively raising the threshold. Thus, the neuron changes as it is used. This provides a mechanism for memory, and learning becomes possible.[2]

---

[1] Generally attributed to Professor Donald Hebb in 1949, who was chair of the psychology department at McGill University at the time. "Hebbian Learning" (Hebb, 2002) is still the term used for various methods of adjusting the weights in artificial neural networks.

[2] From Hebb's 1949 book (Hebb, 2002): "When an axon of cell A is near enough to excite cell B and repeatedly or persistently takes part in firing it, some growth process or metabolic change takes place in one or both cells such that A's efficiency, as one of the cells firing B, is increased." This is more often paraphrased as "Neurons that fire together wire together."

Figure 8.3: Connections between Neurons

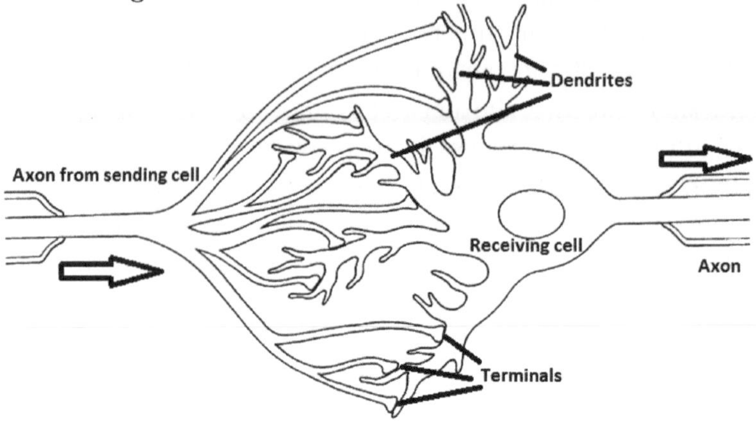

In summary:

1. A neuron

    (a) adds up its inputs,

    (b) tests the total input strength

    (c) creates an output or not depending on the input strength

    (d) adjusts the strengths of the connections

2. A neural network is a goup of connected neurons

    (a) with waves of signals moving through them

    (b) that constantly readjusts its connections,

    (c) thus changing the pattern of signals

    (d) thus allowing our basketball player to learn to score.

This mechanism provided the basis for designing artificial neural networks, and through the 1970s, it was one of the avenues that seemed to hold much promise for artificial intelligence. Unfortunately, that promise was not realized, and by the end of the 1980s research funding for AI began to dry up. Researchers involved started to focus on other applications for artificial neural networks, one of which was predictive modeling.

By treating known historical predictor variables as inputs and known historical target variables as outputs, an artificial neural network implemented as a computer algorithm could learn to provide a prediction for a target by adjusting the strength of the connections between the artificial neurons (Figure 8.4).

Figure 8.4: Comparing Actual and Artificial Neural Networks

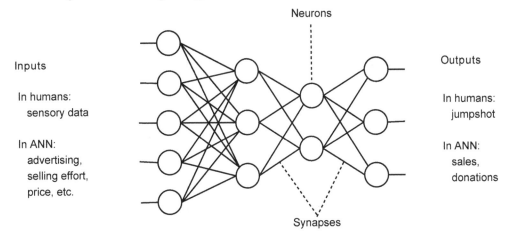

In the ANN, "nodes" take the place of biological neurons. Of course, there is quite a bit of detail that goes into adjusting the strength of the connections between the nodes during training.

In the artificial neural network, there is one *input* node for each predictor variable. This layer of nodes, on the left side of Figure 8.4, is *passive*, meaning that the nodes do nothing except pass the variables to the next layer of nodes. On the right-hand side of Figure 8.4, the *output* layer of nodes gives the predicted value of the target variables. In our examples, we only have a single target variable, but neural networks generally can have more than one output. Between the input and output nodes are *hidden* layers. The figure shows two hidden layers (the first with three nodes, the second with two nodes). In practice, the analyst selects the number hidden layers and the number of nodes in each hidden layer. For most marketing applications, a single hidden layer is adequate, and we will confine ourselves to a single hidden layer architecture in our models. The hidden and output layers contain *active* nodes, which means that this is where the calibration work is done, as described below. In a fully interconnected neural network, the type that we will use, the input nodes pass each variable to all of the hidden layer nodes.

In summary, our neural network models will be fully interconnected, single hidden layer, single output node models. We will have control over the number of nodes in the hidden layer, and of course over the number of input nodes, through the number of predictor variables used.

Figure 8.5: The Algebra of an Active Node in an Artificial Neural Network

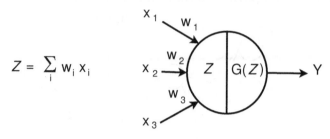

$$Z = \sum_i w_i \, x_i$$

8.2   Artificial Neural Networks as Predictive Models

As with all other types of predictive models, there are many variations on neural network models, and they have been developed and applied in many fields. Exactly how neural networks behave depend on the details of their implementation, which can vary in a number of ways:

1. **Network properties** (the number of nodes, the number of layers in the network, and the order of connections between nodes)

2. **Node properties** (threshold, activation range, transfer function)

3. **System dynamics** (initial weights, learning rule, and weight decay)

These properties are all implemented as mathematical relations between inputs and outputs, and we present a brief outline of one approach.

Each individual node is a model that relates a number of inputs to an output. Figure 8.5 shows the structure of a single node as a combination of a linear model followed by a transformation, G, of the output of the linear model. The linear model is completely analogous to the linear model in regression, with the coefficients called weights and represented by w_i (w_1, w_2, and w_3 in our example). The inputs (predictor or independent variables) are the x_i, and the output (dependent) variable is Z.

$G(Z)$ simulates the threshold behavior of the neuron, and passes the signal along as "1" if Z exceeds a threshold level T, or blocks it by setting the output to zero if Z is less than the threshold T. As shown in Figure 8.6, the transfer function G can be a "hard" transformation, with the output Y only taking the values zero or one. It can also be a soft transformation, with intermediate values allowed via a sigmoidal, or S-shaped, function (like our old friend the logit model). This allows for some "leakage" activation if the value of Z is

Figure 8.6: Hard and Soft Transfer Functions

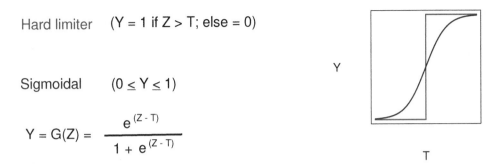

Hard limiter (Y = 1 if Z > T; else = 0)

Sigmoidal $(0 \leq Y \leq 1)$

$$Y = G(Z) = \frac{e^{(Z-T)}}{1 + e^{(Z-T)}}$$

close to the threshold. In this case, the relation between the x inputs and the Y output in our model for a *single node* looks very much like a logistic regression model.

The ANN connects many of these nodes together. One could take any specific architecture, and write out the full mathematical model by substituting the outputs of one set of nodes for the inputs of the following nodes. With the sigmoidal logit transfer function for all nodes, it would look like a complex linked set of logistic regression functions. You can see how this high degree of complexity allows for a lot of variety in the shape of the relationships between the original inputs and the final output of the network. You can also see that the weights, once the model is calibrated, are similar to the coefficients in the logistic regression model. Finally, it should be apparent that interpreting these weights, unlike in the comparatively simple logistic regression model, is essentially impossible. The neural network model sacrifices parsimony for greater completeness.

The calibration process has the same objective as our earlier models, which is to minimize the difference between the predicted target of the model and the observed target in the historical estimation data. The algorithm starts with trial values for the weights, and calculates a predicted output using those trial weights and the input predictor variables. It then compares that predicted output value with the observed actual target value, and calculates the magnitude and direction of the difference, or "error" in the calculation. From this, the algorithm then uses optimization techniques to adjust the weights so as to reduce this error. The process repeats, and continues reducing the error until a stopping rule is activated.

As always, overfitting needs to be controlled for, which is especially important for neural networks given their flexibility. The analyst chooses the number of nodes to put in the hidden layer. The more nodes that are used, the greater the flexibility of the neural network, and the greater the likelihood of overfitting.

Overfitting also tends to occur when some of the weights are very large. To prevent this, a weight decay factor can be used that places a penalty on the error function for large weights, thereby discouraging them. A weight decay parameter of zero indicates no penalty, with larger values of the decay parameter providing an increasing penalty for large weights. Ultimately, we will use a lift chart of the model applied to validation data to select the best model.

In summary:

1. The neural network can "learn" to represent any data set—with any type of functional relationships among the variables—to any required degree of accuracy with a sufficient number of nodes and hidden layers.

2. This allows us to capture underlying relationships without knowing the functional form (structure) of the relationship.

3. And our basketball player, or any other human, doesn't need to know Physics 101 to be able to perform exceedingly difficult tasks. He just has to train a lot.

8.3 Neural Network Models Tutorial

The major advantages of neural networks are that they are extremely flexible and can capture many sorts of relationships. Like decision trees, they easily handle nonlinearities and interactions. Unlike decision trees, they are also very sensitive to weak effects. The major disadvantage of neural network models is that the calibrated output is essentially impossible to interpret. If all we care about is prediction, this is not a problem. If we want to know which predictor variables are related to the target variable, we have to use other predictive models.

The instructions given in this tutorial assume you are starting with an essentially new workspace, requiring you to load the CCS data from the BCA data library. If you are working with the same R workspace you saved after completing previous tutorials, you will need to alter some steps in what follows, and completely skip others.

1.

Read in the CCS data set from the BCA package, and set the Estimation and Validation Samples to 50/50, as before. We will take advantage of our previous explorations of this data set, and, for this tutorial, work with variables which we have found to be predictors of MonthGive. To compare models, we will also regenerate our preferred final logistic regression (MixedCCS2) and tree model (Full011). First, re-create the two-level New.Region variable. Use the menu option **Data → Manipulate variables → Relabel factor levels...** to create the variable New.Region in which regions R2 and R3 are assigned to the level "VanFraser," and regions R1, R4, R5, and R6 are assigned to the level "Other." Then **create the new variables** Log.LastDonAmt and Log.AveDonAmt using **Data → Manipulate variables → Compute new variable....** Recall that these two variables do not have zero values and hence do not require the "+1" term in the log expression. Check what you have done using the **View data set** button.

2.

Select **Model → Statistical models → Generalized Linear Model...** and in the dialog box **select MonthGive as the target, and AveIncEA, DonPerYear, Log.AveDonAmt, Log.LastDonAmt, and New.Region as predictor variables. Set the sample to Sample=="Estimation"and label this model MixedCCS2. Click OK.**

Then select **Model → Machine learning models → Tree Model → Train rpart tree...** and select MonthGive as the target, and all of the remaining variables except the three newly created variables and Sample as predictors (recall that pruning using the cp parameter will automatically eliminate many variables from the final model). Set Sample=="Estimation" and cp = 0.011. Name the model Full011 and click OK to run the model. Select Model → Machine learning models → Tree Model → Plot rpart tree... and click OK.

3.

Select **Model → Machine learning models → Neural network model...** to bring up the dialog box (shown in Figure 8.7). **Set MonthGive as target.** The first exercise will be to observe the effect of increasing the numbers of nodes (called "units" in R) and therefore the flexibility of the model to capture detail in the data. **Select AveIncEA, DonPerYear, AveDonAmt, LastDonAmt, and Region as predictor variables. Set the sample to "Estimation," and the decay parameter slider to 0.10.** This level works well for our purposes and we will always use it. We will explore three different models, with 2, 3, and 4 hidden layer units. For this first model, **set the hidden layer units slider to 2, and name the model NN.HL2. Click on OK.**

We indicated earlier that the number of hidden layers was one of the elements

Figure 8.7: The Neural Net Model Dialog Box

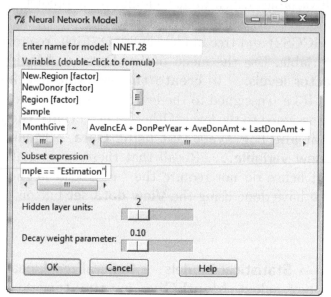

an analyst can alter in a neural network's structure. However, only a single hidden layer is allowed in the underlying R function we are using (which is called **nnet**). At first this may seem like a serious limitation, but, based on our expreience, in marketing applications simpler, single hidden layer neural networks almost always predict new data better than multiple hidden layer networks. While only a single hidden layer is allowed, the number of nodes in that layer can be selected (the number corresponds to the selection of "hidden layer units"). In addition, the analyst can set the decay weight parameter, which we will set to 0.01.

The output window for the neural network model **NN.HL2** is shown in Figure 8.8, and it shows the calibration coefficients (or, as the jargon goes in neural networks, the "training weights") estimated between the five input nodes, the two hidden layer nodes, and the final output or target node. These weights, while inspired by synapse strength, are similar to estimated coefficients in calibrated regression and logistic regression models. The big difference is that it is much more difficult to interpret these weights in terms of a predictor variable's impact on the target variable.

4.

Run two more neural network models with all parameters the same, **except use four hidden layer units in one model, and six hidden layer units in the next model**. The dialog box will return with the variables previously used. Remember to set the decay parameter to 0.10. **Name these**

Figure 8.8: Neural Network Model Results

```
> summary(NNET.28)
a 9-2-1 network with 23 weights
options were - entropy fitting  decay=0.1
 b->h1 i1->h1 i2->h1 i3->h1 i4->h1 i5->h1 i6->h1 i7->h1 i8->h1 i9->h1
  0.22   0.00   1.23  -0.03   0.06   2.62   0.97  -0.27  -0.65  -0.58
 b->h2 i1->h2 i2->h2 i3->h2 i4->h2 i5->h2 i6->h2 i7->h2 i8->h2 i9->h2
  3.19   0.00  -0.96  -0.31  -0.01   3.20  -0.29   0.19  -0.85   0.00
 b->o h1->o h2->o
-1.42  2.89 -2.69
```

two models NN.HL4 and NN.HL6. You will likely notice that with more nodes, the model takes a bit longer to run.

Figure 8.9: Estimation Sample Cumulative Captured Response

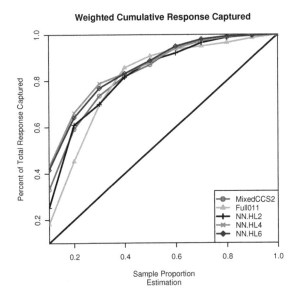

5.

Construct cumulative response lift charts of the five models using the estimation sample (Assess → Graphs → Lift Charts) and select the five models. Recall that the true response rate is **0.01**, the target variable level is **Yes**, and name it **CCS Estimation Sample** to identify the sample used. The result should appear as in Figure 8.9. A model with two nodes gives similar lift to the logistic model, which we put quite a bit of work into to get a good model. At different sample percentages, different models perform slightly differently, but note that increasing the number of nodes generally increases the ability of the model to follow variations in the data, especially over the first few deciles. But wait! Are these "better" models following changes that are repeatable in different data sets, or are they just getting better at tracking unrepeatable noise?

Figure 8.10: Validation Sample Cumulative Captured Response

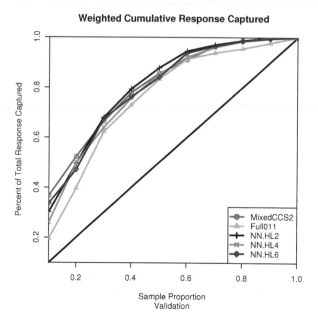

6.

The critical test of the models is to see how well they predict using new data that were not used in calibration. **Create the lift charts of the five models as before, this time selecting the validation sample, and naming the plot CCS Validation Sample.** Now the performance of the neural network models are reversed over much of the data range, indicating that four and six nodes allow too much flexibility, and overfits the estimation data (Figure 8.10). Our best logistic regression model, and the two-node neural network perform the best, especially with the best (first and second decile) prospects.

7.

The neural network may not be able to take full advantage of its flexibility here because we have restricted it to use only the variables that the previous models found to be related to the target. That is, we may be giving our best logistic regression model an unfair advantage. To explore this, **run neural network models with all of the remaining variables used as predictors.** Use the three newly computed variables (**Log.LastDonAmt**, **Log.AveDonAmt**, and **New.Region**) rather than the original variables. Make sure you don't enter the target, MonthGive, or Sample into the predictor list. Try different numbers of nodes. Compare your new models with the `MixedCCS2` model by **generating lift charts on both the estimation and validation data sets. Briefly describe your results.**

Chapter 9

Putting It All Together

At this point you have had an opportunity to gain some experience using the data mining tools contained in R and R Commander. Given this basic background, the primary goal of this chapter is to provide you with a framework that will allow you to rapidly develop a reasonably good model. A model can usually be improved by devoting more effort to it, but there are diminishing marginal returns to this effort. The popular 80/20 rule applies, and the "reasonably good" model is sometimes referred to as the "80% model," implying that with 20% of your modeling time and effort you will get the first 80% of the return, and it will take 80% of your time to get the remaining 20% return (that results in the best possible model).

Before presenting our rapid model development framework, we first present an additional tool, stepwise variable selection (Venables and Ripley, 2002). Stepwise variable selection is an automated way of finding the set of predictor variables in a regression model (either linear or logistic) that is the best based on some criterion (often AIC).

A tutorial showing you how we applied the framework to a new database follows the presentation of the rapid development framework. At the end of the tutorial we will show how to use R Commander to "score" a database. Scoring a database amounts to creating a new variable in a database that allows the user to rank each customer in the database in terms of their attractiveness as a campaign target based on the fitted values of a particular predictive model. We are already familiar with this idea: scoring is exactly what is required to create a lift chart, but the score is not added to the database.

9.1 Stepwise Variable Selection

To illustrate what stepwise variable selection involves, we begin with a simple example. Consider the following simple three-variable linear predictor for either a linear or logistic regression model:

$$a + b_1 \text{Var1} + b_2 \text{Var2} + b_3 \text{Var3},$$

where a, b_1, b_2, and b_3 are coefficients to be estimated, and Var1, Var2, Var3 are predictor variables that can be continuous numeric variables, categorical factors, or a mix of the two. In what follows, this is the linear predictor for the "maximal" model to be considered (which typically does not correspond to a model containing all the variables in a database, but does contain the set of available variables that the analyst thinks might be useful predictors).

The goal of stepwise variable selection is to determine the set of variables out of Var1, Var2, and Var3 that results in the model with the minimum AIC value.[1] There are three possible ways of doing this. The first (known as backward stepwise variable selection) begins by fitting the maximal model to determine its AIC, and then estimate the models that contain the variable sets {Var1, Var2}, {Var1, Var3}, and {Var2, Var3} to determine the AIC values for these three two-variable models. If the AIC is lowest for the maximal model, then the stepwise variable selection process terminates, returning the maximal model. On the other hand, if one or more of the other models (each containing one less variable than the maximal model) has a lower AIC value, the model containing the set of variables with the minimum AIC will be taken to the next step, thereby eliminating one variable from the maximal model. For this illustration, assume that the model with the minimum AIC value includes the variables Var1 and Var3 (dropping Var2). In the second step the AIC for this model will be compared to the two models that contain only a single variable (a model that contains only Var1 and a model that contains only Var3). If the model containing both variables has the minimum AIC, the process terminates returning this variable. However, if one or both of the single variable models has the minimum AIC, the single variable model with the minimum AIC is returned, and the process stops.[2]

The second approach (known as forward stepwise selection) is to compare all possible single variable models, and select the single variable model with the minimum AIC in the first step. For our illustration, assume that the best single variable model contains only variable Var3. In the second step the single variable model that contains Var3 will be compared to two-variable models that contain the set of variables {Var1, Var3} and {Var2, Var3}. Both alternative models contain Var3, so what the algorithm is really doing is determining the next best variable to add to the model. If the single variable model containing Var3 has the minimum AIC, the process terminates, and this model is returned. If one or both of the two-variable models has a lower AIC than the single variable model, then the two-variable model with the smallest AIC

---

[1]In this discussion we focus on the AIC, but criteria other than the AIC can be substituted.

[2]The process could continue comparing the best single variable to a constant-only model, but most stepwise variable selection procedures take as the "minimal" model one that contains a single variable.

is moved to the next step; assume that this is the model containing Var1 and Var3. In the next step the AIC of a model containing Var1 and Var3 is compared to the AIC for the model containing all three variables (the maximal model). If the maximal model has the smallest AIC it is returned, otherwise, the two-variable model containing Var1 and Var3 is returned.

The final approach combines both the backward and forward stepwise approaches. Specifically, this approach (like backward stepwise selection) begins with the maximal model and determines which (if any) variable should be removed to obtain the lowest AIC possible. In our example this would involve removing Var2 from the maximal model. In the second step, the model containing Var1 and Var3 is compared with models containing each of these variables by themselves (alternatively dropping Var1 and Var3) and with the model containing all three variables (adding Var2). Assume that the model with the lowest AIC contains only Var3. In the final step, this model is compared with the two models containing the set of variables {Var1, Var3} and {Var2, Var3}, and the model with the lowest AIC is returned.

In our three-variable maximal model example, the final approach results in a number of redundant tests, and would provide the same final set of variables as either the forward or backward stepwise selection approaches. However, as the number of variables increases, the final variables selected could differ between stepwise regression models. In backward variable selection, once a variable is removed, it will never be included in the set of variables in a subsequent step, while in forward variable selection, once a variable is added, it will never be removed in a subsequent step. In contrast, the combined approach allows a variable that has been added in one step to be removed in a subsequent step and vice versa. As a result, it selects the set of variables from the maximal model that has the smallest AIC possible, but at the cost of doing redundant comparisons. While this is inefficient, with modern computing power it can typically be done in a reasonable amount of time.

While stepwise regression can find the subset of variables contained in the maximal model that has the lowest AIC, it does not transform the variables in the maximal model. As a result, a variable may be rejected by stepwise variable selection methods, but the natural logarithm of that variable could be included in the final set of variables selected by these same methods by making appropriate changes to the maximal model. In a similar way, a categorical factor may be removed, while a version of this variable with a reduced number of levels may be accepted.

9.2   The Rapid Model Development Framework

The basic steps in the framework are:

1. Think about the behavior that you are trying to predict.

2. Carefully examine the variables contained in the data set.

3. Use tree and regression models to find the important predictor variables.

4. Use a neural network to examine whether nonlinear relationships are present.

5. If the neural network gives better lift (particularly for the validation sample), use visualization tools (those shown in the tutorials to Chapters 4 and 5) to look for and better understand nonlinearities. If not, stop.

6. Try to improve the lift from the regression model by transforming variables and including them as inputs (for example, create a new variable that is the square or log of a variable that appears to have a nonlinear relation with the target variable).

Having given you the steps, we next introduce the database that is the subject of the tutorial to provide a concrete example of an application, and to explore the framework in more detail before applying it.

9.2.1   Up-Selling Using the Wesbrook Database

The database that you are about to apply the rapid development framework to contains a list of donors provided by the UBC Development Office. The UBC Development Office is responsible for all fund-raising activities at the University of British Columbia. These data come from a data mining project that was conducted in an effort to find ways to increase the efficiency of UBC's fund-raising activities. With this data set, the goal is to identify current UBC donors who have a profile that is similar to a "Wesbrook" donor's profile. A donor is assigned to the Wesbrook class if that individual has a donation that exceeded $1000 in a previous year. The Development Office actively pursues donors that it identifies as potential Wesbrook-level donors in an effort to demonstrate to them the benefits of donating at the Wesbrook level. This is a classic "up-selling" exercise, and one of the most common applications of data mining techniques. The basic premise is that you identify those people in your database who have not yet donated at the Wesbrook level, but have a profile similar to those that have donated at the Wesbrook level. Targeting those with higher likelihoods of being larger donors should result in a better response rate for UBC's marketing activities.

9.2.2 Think about the Behavior That You Are Trying to Predict

We know that the best Decision Support models are based on good theoretical models. While it is unlikely that you will want to develop watertight theory, developing a mental model of the problem will help frame your subsequent model-building efforts. In the case of a Wesbrook donor, what factors do you (based on your own beliefs) feel are likely to influence whether someone will give $1000 or more to UBC in a single year? For instance, you may believe that (1) an individual's ability to make a fairly large donation in a single year and (2) the individual's personal feelings toward UBC are the most critical factors in determining whether someone would donate at the Wesbrook level. Ultimately, you want to develop an initial mental model of how the underlying process works so you will be in a better position to critically evaluate the results provided by the data mining tools.

At some point you may find yourself dealing with a project in a domain with which you are unfamiliar, but you have a client that is very knowledgeable. In this instance, it is worthwhile spending time with that client determining what that client's mental model is in order to determine which possible predictor variables are likely to matter most, and what is the expected nature of the effect of a variable (i.e., what does the client think will happen if one variable increases in level with respect to the probability that a customer takes the desired action). Knowing this in advance can point out potential problems (likely with the data) that can undermine your relationship with the client.

9.2.3 Carefully Examine the Variables Contained in the Data Set

There are four reasons to carefully examine the variables in your data set. The first is to determine what measures are available in the data set that are related to the factors you (or your client) believe should be important in predicting the behavior of interest. Some of the measures will be readily available in the data set, while you may be able to construct other measures out of the available data.

The second reason to closely examine the data is to determine which variables are likely to exhibit missing value problems. The more variables with missing values that you use, the more data you are going to have to reject (unless each variable has its missing data for the same donors, as is the case for many of the variables associated with donors without UBC degrees). If 10% of the values of a variable consist of missing values, you will have problems. Anything more than 3% should at least be thought about. Both of these numbers are somewhat arbitrary, and are only intended as a rough guide. If you do have variables with missing value problems, you will need to address them prior to conducting regression or neural network modeling. How you deal with missing values is context specific, and this issue will be highlighted in the application of the framework in this chapter's tutorial. However, early on you need to determine whether a missing value for a variable is due to factors related to the construction of the database itself or if it is a result of non-reporting issues (e.g., an individual simply did not provide some information). Knowing the "cause" of a missing value can help in deciding what to do about missing instances of that variable.

Similar to the last reason, the third reason to closely inspect the data is to find categorical factor variables that have levels that contain a small number of records. A number of problems can arise from this issue. Specifically, the statistical reliability for the indicator variables generated for these factor levels is very low. This situation can also cause problems with certain tools such as the plot of means visualization method. Finally, if multiple samples are created within the database (i.e., estimation, validation, and holdout samples), certain samples may contain no records that possess a particular level of a factor, which creates problems in predicting the target variable for samples other than the estimation sample, causing the lift chart tools to fail. The way to deal with these problems is to create a new factor with a smaller number of levels (and with more records in "thin" categories) from the original factor.

The fourth thing you need to think about is whether there are variables in the database that are either used to construct the target variable of interest or are "trivially related" to the target variable. As an example of a variable that is "trivially related" to the target variable, consider a variable that indicates whether an individual has purchased an extended service plan in a model that predicts whether an individual will purchase a major appliance, using department store transaction data. The variable is likely to be strongly related to appliance purchase, and hence appear as a very good "predictor" of whether someone buys an appliance. However, if we consider the timing of the two variables, it is easy to see that this would be due to a reverse causality. Specifically, purchasing an extended service plan occurs *after* purchasing a major appliance, so it cannot cause the appliance purchase — the appliance purchase causes the service plan purchase. Including such predictors in the model will often give very good fit and lift but will be of little value when it comes to identifying new customers' purchase likelihood. Therfore, variables like this should never be used as predictors, once again highlighting the critical importance of understanding exactly what all of your variables represent in the real world.

Finally, you need to set the role of database housekeeping variables to either a case name identifier or remove them. Housekeeping variables are used to sort and merge records, but are not relevant in explaining underlying behavior. Sometimes an id number will come up among the set of predictors in a stepwise regression since the lower the id number, the longer the individual has been a part of the database, and thus the older they are likely to be. In this instance the id variable is acting as what is known as a "proxy variable" for age, but it makes more sense (and is much more interpretable) to directly include the age variable instead. At other times, the id variable is capturing random patterns in the data that will simply not be present in new data. This can be handled by methods to control overfitting. Postal and ZIP codes can also be a problem. If they are treated as categorical, and there are hundreds, the regression model will add hundreds of indicator variables, and many may not be estimable. If they are treated as interval (as U.S. five-digit ZIP codes might be) they might capture some difficult to interpret locational factors (e.g., U.S. ZIP codes increase in numeric size, say from 01111 to 99999, from east to west across the country). Appropriate geographic and geo-demographic variables should be included rather than the postal code itself.

9.2.4 Use Decision Trees and Regression to Find the Important Predictor Variables

Finding the set of important predictor variables is about 90% of the battle in developing a good predictive model. Using decision trees and regression can often help you quickly find the most promising set of predictor variables. Each of these methods has its relative strengths and weaknesses in finding the set of important predictor variables, but when both are used, you can take advantage of the strengths that each offers. Specifically, decision trees are often good at finding the most important predictor variables, and they seem to be less likely to find variables that are simply correlated with random patterns in the estimation sample. In addition, tree models are capable of detecting predictor variables that have non-linear effects on the target. As an example, consider the influence of a person's age in a model that is attempting to predict how much an individual will invest in mutual funds in a particular year. It is likely to be the case that mutual fund investments will start at a low level for the typical individual when she is young, will increase each year until this individual is in her late 50s or early 60s, and then will begin to decline (perhaps rapidly) as she enters her senior years. In this instance, two two-way splits for age would indicate relatively low levels of new mutual fund investments for the youngest and oldest groups, and a higher level for a middle group.

The strength of regression is its ability to find those weaker, but still important, predictor variables. However, finding these weaker variables comes at the cost of finding more variables that are included only to capture random patterns in the data. Another relative weakness of regression is that it can "miss" variables that have a non-linear relationship to the target if potential nonlinearities are not explicitly taken into account. In our mutual fund example, the age variable is unlikely to be found by regression. The reason is that the relationship between age and mutual fund investing looks like an inverted U. As a result, throughout the entire range of age, the best-fitting line would have a slope of zero. Although, if you broke the age range into two variables (one that gives the age of someone aged 59 and below, and the other giving the age of someone aged 60 and above) you would likely find a strong positive relationship between age and mutual fund investing for the first variable, and a strong negative relationship for the second. Unfortunately, you would need to know that you should split up the variable this way prior to running the regression models.

Having said all this about the different tools, what we are really getting at in this step is the interplay between your initial mental model of customer behavior and the data. Specifically, your mental model will likely be incomplete, and the two modeling tools will point you to other variables that you will feel comfortable incorporating into your mental model. However, the tools may also point to other variables that simply do **not** make sense given your mental model. If this is the case, there is a very good chance that these variables actually do not explain the behavior, and were included as a result of a particular model's attempt to account for random patterns in the data. If you estimate a model that includes variables you do not think make sense, remove these variables and re-run the model.

9.2.5 Use a Neural Network to Examine Whether Nonlinear Relationships Are Present

Neural networks are very good at finding and mimicking nonlinear patterns in the data. Because an estimated neural network model is nearly impossible to interpret, they are not good at finding the set of predictor variables to include in the model. However, in the prior step of the framework you have hopefully uncovered the important set of predictor variables. Moreover, through the use of a decision tree, you have had a preliminary opportunity to see whether any important variables appear to have a nonlinear relationship with the target. The use of a neural network will allow you to assess the extent to which nonlinear relationships in the data matter for purposes of prediction. The way that you accomplish this is by first estimating a fairly flexible neural network (one with three hidden nodes in the hidden layer). After you have estimated the network, compare it to your final stepwise regression model through the use of lift charts that examine the models performance on the estimation and validation sample. If your final stepwise regression model performs about as well as or better than the neural network across the samples, then you should view your final regression model as the "good enough" (or "80%") model. If the neural network outperforms the regression model in the validation sample, *and* it is important to know the details of the relation between predictors and target, then it is time to try and build a better regression model that captures the nonlinear patterns in the data.

9.2.6 If There Are Nonlinear Relationships, Use Visualization to Find and Understand Them

Investigating nonlinear relationships is only necessary for continuous *predictor* variables. If the *target* variable is continuous, then looking for nonlinear relationships can typically be accomplished by looking at a scatterplot and/or line plot of the predictor and the target variables. If the target variable is binary, and the predictor variable under investigation has an ordinal categorical scale, then a plot of means of the target (converted to 0–1 numeric) that uses the predictor variable as the grouping variable in the plot works well. The most difficult relationship to examine graphically is one that involves a binary target variable and a continuous predictor variable. The best way to look for any potential patterns is to use the methods (e.g., binning the predictor) we present in the logistic regression tutorial of Chapter 5. In general, if you cannot visualize an obvious relationship between a predictor and the target, the nonlinearities are not likely to be strong enough to matter much in the model. If you see a decreasing returns concave type of relation, a logarithmic transform of the predictor should help. Replace the predictor with its logarithm in the regression model and check the lift on the validation data. A U-shape relation can be captured by a squared predictor (remember to include both the original variable *and* the square of that variable). How far you wish to go in this search for improvement depends on how much time you have, keeping in mind that the incremental improvements you get will be smaller and smaller as you proceed.

9.3   Applying the Rapid Development Framework Tutorial

A special note for what follows. The main steps in the tutorial are in a bold typeface, with explanations given in a regular typeface.

This is a good place for a reminder that it is absolutely essential for the analyst to understand the meaning and source of all variables, whether they come from surveys, transactional data, census data, or anything else.

Database construction will typically include oversampling of the target variable, and then separating the sample into an estimation (or "calibration"), validation (or "holdout"), and (possibly) holdout samples. For this exercise, the database was oversampled so as to have a 50/50 split of Wesbrook/ non-Wesbrook donors. The database was then exported to a dBase format file for easy importation into R.[3] The name of the dBase format file is Wesbrook.dbf, and it contains 2770 records. You can download a zip archive containing the data file using the link http://www.customeranalyticsbook.com/ Wesbrook.zip.

Import the Wesbrook dBase file into R, (Data → Get from → dBase (dbf) data set. . .), naming the dataset Wesbrook and UNchecking the "Do not convert character variables to factors" **checkbox,** since we do want the character variables to be factors (categorical). When done, the bottom of the R Commander window should indicate that you have imported 2770 rows and 31 columns.

Given the objectives of this tutorial, we will only create estimation and validation samples, and no holdout sample. Moreover, given the fairly small size of this data set, we will place 70% of the records into the estimation sample, and 30% into the validation sample. To do this, **use the pull-down menu option Data → Organize → Create samples in active data set. . . , and allocate 70% of the observations into the estimation sample and 30% into the validation sample. Check your results with View data set.**

---

[3]This format belongs to a database management product called dBase that was the dominant PC database management product during the early days of personal computing (it pre-dates the introduction of MS-DOS, let alone Windows). While dBase is no longer dominant, a number of popular products use the file format. The reasons for this is that the files are fairly small (though not as small as a tab-delimited or comma-separated value text file), compress well, developers of other software packages find them easy to work with (which is not true of Excel files), and they contain enough information about the variables within the file that the variable types are correctly imported (which can be a real issue with either tab-delimited or comma-separated value text files, and even directly imported Excel files). It is the presence of the variable descriptors that make dBase files (which have the suffix of *.dbf) slightly larger than tab-delimited or comma-separated value text files.

Step 1: Mental Model

As is hinted at above, our mental model is based on the notion that there are two important determinants of whether an individual will donate $1000 or more to UBC in a single year, specifically, how favorably someone views UBC based on her/his personal experience with the university, and on her/his ability to afford a fairly large donation in a single year. We know that measures of her/his ability to afford a fairly large donation will be easier to obtain than measures of her/his personal experiences with UBC. However, even measures related to this second factor are likely to be available in the database.

Step 2: Examine and Clean Data

Are there measures related to our mental model? Start by looking at the available measures in the data (excluding the `Sample` variable we just created), which are shown below. Looking at the available data, several measures of the ability to afford a large contribution are readily available in the database (Table 9.1). "EA" refers to a census "Enumeration Area" and identifies neighborhood variables.

Specifically, the number of years an individual has been out of school, and the faculty from which he or she graduated, should be related to his or her ability to make a large contribution. In addition, the average income of households and the average value of dwellings (a measure of both income and wealth) for the enumeration area in which the individual resides should also provide good measures of an individual's ability to make a fairly large contribution. Again, remember that these demographic variables are actually geo-demographics, and, hence, give the characteristics of the donor's neighborhood, not the individual donor.

In terms of potential measures related to an individual's feelings toward UBC, there are a few. Specifically, the faculty and/or department an individual graduated from is likely to influence her/his feelings toward the university. Another potential measure of an individual's feelings may be related to whether other members of her or his family (i.e., parents, children, and/or spouse) also attended UBC. All of the above measures are readily available in the database.

Two additional variables that do not add information, but are easier to interpret at the end, will be created: `YRFDGR` (the years since the individual received his or her first UBC degree: 1999 – `FRSTYEAR`) and `YRLDGR` (the years since the individual received his or her most recent degree from UBC: 1999 – `GRADYR1`).[4] These variables will be easier to interpret than the year of grad-

---

[4]Note that the label of `GRADYR1` is misleading — it says "1st UBC graduation year." However, they are counting from the present backward to the past, so 1 is actually "latest."

Table 9.1: The Variables in the Wesbrook Database

Variable	Label
ATHLTCS	Participation in athletics
AVE_INC	Average 1995 household income in the EA of residence
BIGBLOCK	Participation in Big Block
CHILD	Child of UBC student or alumnus code
CNDN_PCT	% Canadian citizenship in the EA of residence
DEPT1	Department of 1st UBC degree code
DWEL_VAL	Average value of dwellings in the EA of residence
EA	Enumeration Area universal ID
ENG_PCT	% English as home lanugauge in the EA of residence
FACSTAFF	UBC faculty or staff member
FACULTY1	Faculty of 1st UBC degree code
FRSTYEAR	Earliest UBC graduation year
GRADYR1	1st UBC graduation year
HH_1PER	1 person households % in the EA of residence
HH_2PER	2 person households % in the EA of residence
HH_3PER	3 person households % in the EA of residence
HH_45PER	4-5 person households % in the EA of residence
ID	Viking ID code
INDUPDT	Date of last update of personal data
MAJOR1	Degree of 1st UBC degree code
MARITAL	Marital status code
MOV_DWEL	% of households in movable dwelling in the EA of residence
OTHERACT	Participation in other activities
OWN_PCT	% of households that own their dwelling in the EA of residence
PARENT	Parent of UBC student or alumnus code
PROV	Two character province/territory abbreviation
SD_INC	Standard deviation of household income in the EA of residence
SEX	Gender code
SPOUSE	Spouse of UBC student or alumnus code
TOTLGIVE	Total lifetime giving
WESBROOK	Ever given at the Wesbrook level

Figure 9.1: Computing the YRFDGR Variable

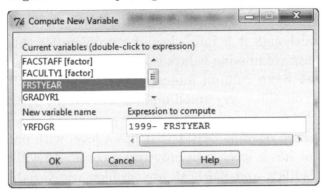

Figure 9.2: The Delete Variable Dialog

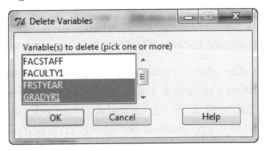

uation variables. **Select Data → Manipulate variables → Compute new variable...** to bring up the dialog box in Figure 9.1. The box is shown with the information for YRFDGR filled. **Click OK. Repeat for the new variable YRLDGR (1999 - GRADYR1).**

You can view the data set to ensure the new variables have been created.

Since we want to make sure and use the two new "years since graduation" variables instead of the original "year of graduation" variables we remove the originals from the data. **Select Data → Manage variables in active data set → Delete variables from data set...** and highlight FRSTYEAR and GRADYR1 (Figure 9.2). **Click OK.**

Look for potential problems with missing values and factor levels with few records.

The next step is to check for missing values by **selecting Data → Clean → Summarize variables**. The results are shown in Table 9.3. The critical columns in the results for our purposes are %.NA (the percentage of missing values for each variable) and Min.Level.Size (the number of records in the smallest level for each factor). An examination of the %.NA column indicates that a number of variables have no missing values, while others have a high percentage of missing values (MAJOR1 has nearly 71% missing values). The particularly problematic variables in terms of missing values are FACULTY1, DEPT1, MAJOR1, MARITAL, YRFDGR, AND YRLDGR, which all have between 31% and 71% missing values. In addition, all the geo-demographic variables have missing values, although it is much less of a problem for this set of variables since the percentage of missing values is always less than 3%. An examination of the Min.Level.Size column indicates that a number of factor variables have levels that contain a very small number of records. Specifically, INDUPDT and MAJOR1 both have a level that contains a single record, PROV that has a level with only three records, DEPT1 that has a level with only seven records, and MARITAL that has a level with only two records. As will be discussed below, INDUPDT, MAJOR1, and MARITAL are problematic variables on a number of dimensions. DEPT1 is also challenging, and has a great deal of overlapping information with FACULTY1, which we also discuss below. This leaves PROV, which is the province/territory indicator of a donor's residence. The natural way to deal with this variable is to group provinces and territories into geographic groups, thereby increasing the number of records in each level.

Carefully inspecting the data summaries shows another data problem; the factor BIGBLOCK only has a single level, which you can see by examining the Levels column in Table 9.3. Such variables (called *unary*) are common in databases. However, they are useless for predictive purposes, and can foul up modeling routines if left in. Any variables with a single value should be deleted. **Select Data → Manipulate variables → Delete variables from data set...**, select BIGBLOCK, and OK.

If someone did not graduate from UBC, then YRFDGR and YRLDGR will have missing values. In this instance missing values are actually due to choices made in database construction and are meaningful, in that they tell us something useful about the individual's relationship to UBC — in this case that they did not receive a degree from UBC (likely they are donors who did not even attend UBC). Similarly, FACULTY1, DEPT1, and MAJOR1, which represent the faculty, department, and major of the individual's first degree, are also meaningful indicators of no degree. However, MAJOR1 has over twice the number of missing values as the other two variables. A visual examination of these three variables (**View data set**) indicates that only certain faculties have majors. For example, it appears that Arts and Applied Sciences do have majors, while Commerce and Law do not. Sorting out how to handle MAJOR1 is likely to be a big job, so we will not use it for our good enough model. However, we should keep it in mind if it turns out we have lots of time to work on it.

Dealing with missing values for faculty and department is much less complex. Specifically, for these two variables, we can simply change the missing values to a new value that indicates that the category for that individual was

Table 9.2: Wesbrook Variable Summary

	Class	%.NA	Levels	Min.Level.Size	Mean	SD
ID	numeric	0.00	NA	NA	1.13e+05	6.71e+04
WESBROOK	factor	0.00	2	1385	NA	NA
TOTLGIVE	numeric	0.00	NA	NA	3.33e+03	7.64e+03
PARENT	factor	0.00	2	284	NA	NA
CHILD	factor	0.00	2	79	NA	NA
SPOUSE	factor	0.00	2	1037	NA	NA
SEX	factor	0.00	2	1081	NA	NA
FACSTAFF	factor	0.00	2	263	NA	NA
ATHLTCS	factor	0.00	2	179	NA	NA
BIGBLOCK	factor	0.00	1	2770	NA	NA
OTHERACT	factor	0.00	2	266	NA	NA
Sample	character	0.00	NA	NA	NA	NA
INDUPDT	factor	0.43	974	1	NA	NA
EA	numeric	0.47	NA	NA	5.66e+07	7.48e+06
PROV	factor	0.61	11	3	NA	NA
MOV_DWEL	numeric	1.01	NA	NA	3.66e-03	3.25e-02
HH_1PER	numeric	1.01	NA	NA	2.26e-01	1.67e-01
HH_2PER	numeric	1.01	NA	NA	3.24e-01	8.67e-02
HH_3PER	numeric	1.01	NA	NA	1.51e-01	5.74e-02
HH_45PER	numeric	1.01	NA	NA	2.56e-01	1.38e-01
DWEL_VAL	numeric	1.01	NA	NA	4.43e+05	2.86e+05
ENG_PCT	numeric	1.01	NA	NA	8.12e-01	1.66e-01
OWN_PCT	numeric	1.01	NA	NA	7.05e-01	2.38e-01
CNDN_PCT	numeric	1.05	NA	NA	8.82e-01	7.96e-02
AVE_INC	numeric	2.60	NA	NA	7.85e+04	3.71e+04
SD_INC	numeric	2.60	NA	NA	1.50e+05	1.48e+05
FACULTY1	factor	34.19	14	19	NA	NA
DEPT1	factor	34.19	23	7	NA	NA
YRFDGR	numeric	34.19	NA	NA	2.67e+01	1.65e+01
YRLDGR	numeric	34.19	NA	NA	2.49e+01	1.65e+01
MARITAL	factor	65.52	6	2	NA	NA
MAJOR1	factor	70.90	142	1	NA	NA

Figure 9.3: Recoding YRFDGR

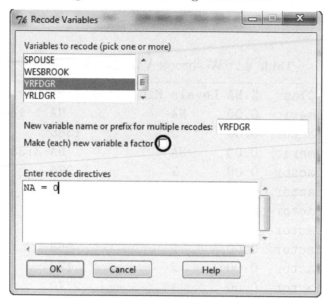

"none" or "no degree."[5] The two "years since ... degree" variables are a bit more difficult to handle since it is difficult to assign a year to a degree an individual did not receive. A common way of dealing with this situation is to force the missing values of each of these two continuous variables to zero, and then add a new indicator variable that indicates that the individual did not receive a degree to adjust for the imputed zero values. In this instance, the indicator variable is likely to be redundant information since if a faculty and/or department variable enters the regression model, then the indicator variable will always equal one when the faculty/department variable equals the "no degree" category. As a result, it will serve as the indicator, and we do not need to create a separate "no degree" indicator variable for years of first and last degrees.

Select **Select Data → Manipulate variables → Recode variable....** In the dialog box (Figure 9.3) **select YRFDGR, and keep the new variable name the same, YRFDGR.** As long as no mistakes are made, this will save some time as this will overwrite the old variable and we will not have to go back and delete the old variable. We want the new variable to be continuous, not categorical, so **make sure to UNCHECK THE FACTOR** box or you will be redoing the variable creation step! The recode directive is **NA=0.** Click **OK**, and agree to overwrite the variable when requested.

---

[5]Remember that if we leave the "NA" code in, R will read that as "missing," and remove the entire case in procedures like logistic regression.

Table 9.3: The New Year since Degree Variables

	Class	%.NA	Levels	Min.Level.Size	Mean	SD
YRFDGR	numeric	0.00	NA	NA	1.75e+01	1.84e+01
YRLDGR	numeric	0.00	NA	NA	1.64e+01	1.78e+01

Figure 9.4: Recoding DEPT1

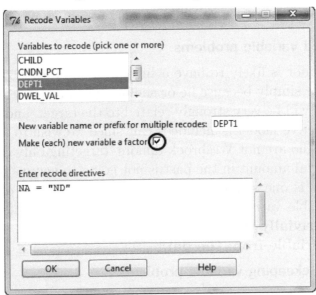

Repeat the recoding for YRLDGR. **View the data to make sure THE "0" replaced the "NA." Also, summarize the data again (Data → Clean → Summarize variables) to see if the appropriate changes were made.** The result should look similar to Table 9.3 — showing mean and standard deviation rather than the levels and minimum size level of a factor, and no missing data.

For DEPT1, which is categorical, recode the value NA to "ND" for "No Degree." Leave the factor box checked this time. Note that when we are recoding to names like "ND" rather than numbers like "0," we put quotation marks around the recode value to indicate that these are character values. The recode dialog box is shown in Figure 9.4. **Again since we kept the same variable name, it is important not to make a mistake!** When it is set, **click OK and agree to the overwrite query.**

Repeat for FACULTY1, setting NA to "ND." Check the data to ensure the replacements have been made.

Missing values for MARITAL are due to non-reporting. They tell us nothing about marital status, only that the individual's marital status was unknown to the UBC Development Office (the marital status of couples where both spouses went to UBC is more likely to be known by the development office). We don't have any reason to think that it will be a useful predictor, nor can we think of any compelling reason that individuals who do not report marital status would have any relation to Wesbrook-level donating behavior, so there is little to gain from treating non-reporting as a valid response. Therefore we will remove it from the data. **Select Data → Manipulate variables → Delete variables from data set...** and select `MARITAL` from the variable list, Click OK.

Trivially related variable problems.

A Wesbrook donor is likely to have a higher value of `TOTLGIVE` than a non-Wesbrook donor simply because he or she has given at least $1000 in a single year. `TOTLGIVE` will be very strongly related to the target, and will likely dominate our predictive model if included. But since we wish to apply our model to individuals who are not Wesbrook donors, targeting only donors who have given a large total amount in the past is not going to help us in our *up-selling* campaign. This is one striking example of the importance of understanding what the available variables mean in relation to your managerial problem. **`TOTLGIVE` is trivially related to the target variable. Consequently, delete this variable from the data.**

Database housekeeping variable problems.

The variable ID is the record-keeping identification number of each person in the database, while EA is a code that identifies the census Enumeration Area in which an individual resides, and has the same properties as a postal code for our purposes, that is, it is not useful. The update variable INDUPDT is not only an internal variable and irrelevant to the individual donors, but also would be treated as a categorical variable with a huge number of categories, so we want to be sure it does not enter any models. **As a result, select Data → Clean → Set record names... and choose ID in the window. Click OK. Also, delete the variables EA,INDUPDT, and MAJOR1 so that they are not accidentally included in a later model.**

As a final check, **View the data set** and **Variable summary for the active data set** to check that all of your changes have been made.

Step 3: Explore with Trees and Regression

We begin the third step of the framework by using a decision tree (**Model → Machine learning models→Tree model → Train rpart tree...**) to see what variables this method finds predictive of Wesbrook donors. **The target variable is `Wesbrook`.** Since we have removed most of the undesirable variables,

Figure 9.5: Estimating a Decision Tree Model

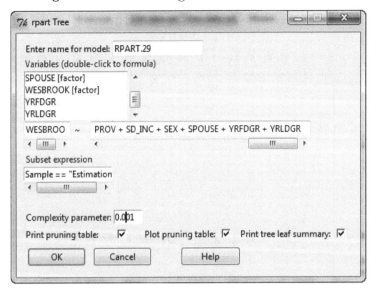

select all of the remaining variables as predictors except `Sample` in the model. The "Subset expression" should be set to `Sample == "Estimation."` Trying the default value for cp of 0.01 indicates that the validation error is still decreasing. Therefore, bring up the rpart tree dialog box again and **set cp = 0.001** (Figure 9.5). The model variables should remain the same. The cross-validation plot for pruning should look something like Figure 9.6 (it will vary because the automatic generation of holdout samples to calculate the misfit on the holdout data varies with each run). From the chart, and based on the criterion of minimum cross-validation error, we should prune the tree back to about 10 or 12 splits, corresponding to a cp value of 0.0072. **Bring up the rpart tree dialog box again,** which should still have all of the model details in it, **and change cp to 0.0072. This time, name the model** `WesTree`. **Click OK to Run the tree.**

Inspect the tree to see which variables were selected **(Model → Machine learning models→Tree model →Plot rpart tree...**) by the tree algorithm (Figure 9.7). These are DEPT1, YRFDGR, DWEL_VAL, SD_INC, AVE_INC, YRLDGR, and FACSTAFF. Happily, most variables fit our prior theory that the ability to donate and feelings toward the faculty of graduation should be relevant. The somewhat surprising variable is SD_INC. However, the standard deviation of income in an enumeration area increases as the average income increases, and is therefore likely a proxy for average income.

At this point we are nearly ready to run and compare logistic regression models. However, logistic regression models omit records that contain missing values for any of the variables included in the model. As a result, if two models

Figure 9.6: The Wesbrook Pruning Plot

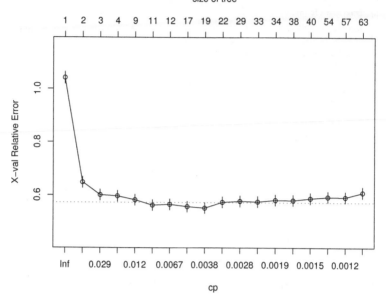

Figure 9.7: The WesTree Model Tree Diagram

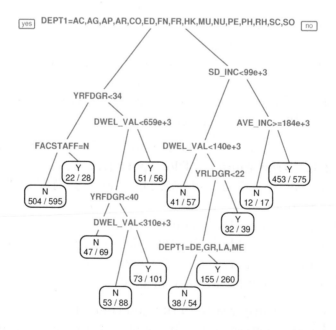

Figure 9.8: Remove Missing Data Dialog

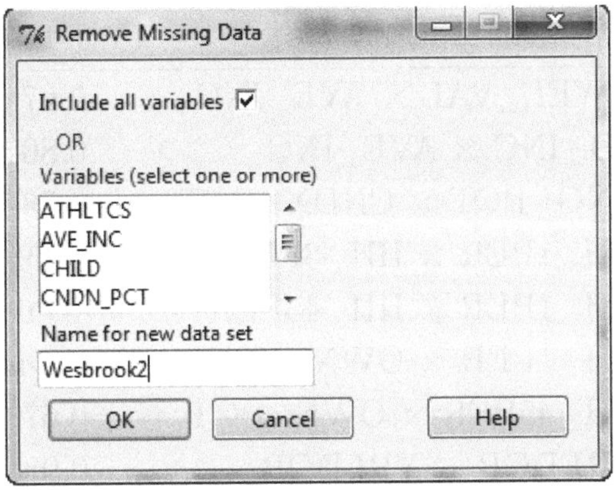

are compared that have slightly different variables, it is possible they will be based on a slightly different set of records based on differences in missing value patterns across variables used in the two models.[6] To guarantee that we are working with the same set of records when examining different models, we want to limit the database to only those records that have no missing values for the variables we are investigating, and then run all models on the same data set. Fortunately, we have done a lot to reduce the number of missing values in the database so far. To create the new data set with no missing values, **select Data → Clean → Remove records with missing data...**, which brings up the dialog box shown in Figure 9.8. In the dialog box **check the "Include all variables" box, and make the "Name for new data set"** **Wesbrook2**. After **clicking OK**, the Wesbrook2 data set will be created and will become the active data set. By only including complete records for the variables of interest, the size of the database has only been reduced by 72 records (or a little under 2.6%). This is a reasonable reduction in the size of the database.

Now **rerun the tree model using the Wesbrook2 data set, setting cp = 0.0072 as before, and name the new model WesTree2. Plot the tree and check that the small reduction in the number of records has not changed the structure** beyond some minor changes in the split points.

The last thing we want to check before beginning the logistic regression analysis is to check the correlation matrix to determine which variable pairs might suffer from collinearity issues. Recall that since correlated predictor variables can appear nonsignificant even if they are both strong predictors, we will want to check for this after we have run the model. In preparation, let's find the

---

[6]Another, related, reason for limiting ourselves to a data set with no missing values in the variables is that the stepwise variable selection tools complain bitterly when different regression models in a step have a differing number of observations.

Table 9.4: Highly Correlated Variables in the Wesbrook Database

Variables	Correlation
DWEL_VAL × AVE_INC	0.67
SD_INC × AVE_INC	0.80
ENG_PCT × CNDN_PCT	0.73
HH_1PER × HH_3PER	-0.69
HH_1PER × HH_45PER	-0.86
HH_1PER × OWN_PCT	-0.74
HH_45PER × OWN_PCT	0.67
YRFDGR × YRLDGR	0.98

strongly correlated variables. **Select Explore and Test → Summarize → Correlation Matrix...** and in the dialog box, select all of the variables. **Click OK**. We scanned through the resulting matrix for numbers around 0.65 or greater (in absolute value) and found those shown in Table 9.4.

Now we are ready to run the "maximal" logistic regression model we will use as the basis for stepwise variable selection. To do this **select Model → Statistical models → Generalized Linear Model**. Select WESBROOK as the target, all of the remaining variables except Sample as predictors, set Sample == "Estimation" as the "Subset expression," and name the model WesLogis. The output of doing this is shown in Table 9.5.

Based on a McFadden R^2 of 0.31, the model appears to fit well. Significant variables are ATHLTCS, AVE_INC, CHILD, several of the DEPT1 indicator variables, FACSTAFF, DWEL_VAL, HH_1PER, HH_3PER, the Quebec level of PROV, SD_INC, SEX, and YRFDGR. Given the level of correlation we saw between some of the variables previously, the statistical significance for a number of them may be greater than what their z-tests suggest. YRFDGR and YRLDGR seem particularly prone to this possible problem. The "NA" values in the FACULTY1 variables are occurring because DEPT1 contains identical information as FACULTY1 for these levels, and they cannot be estimated (technically, due to "singularities").

Now that we have a maximal model (and with 23 variables, it really is a *maximal* model), we are ready to find the subset of its variables that minimize the AIC. Start the stepwise variable selection process by selecting **Assess → Stepwise model selection...**, which will bring up the dialog box shown in Figure 9.9. The only setting you want to change from the defaults is to **enter WesStep in the "Enter the name of the new model:" field**. Once you have done this, **press OK**. At this point it is time to pour yourself the hot or cold beverage of your choice, call someone on your cell for a chat, or find some

Table 9.5: Logistic Regression Results for the "Maximal" Model

Coefficients: (4 not defined because of singularities)				
	Estimate	Std. Error	z value	Pr(z)
(Intercept)	-9.069e+00	2.724e+00	-3.329	0.00087 ***
ATHLTCS[T.Y]	6.679e-01	2.641e-01	2.529	0.01143 *
AVE_INC	1.234e-05	4.696e-06	2.629	0.00857 **
CHILD[T.Y]	1.782e+00	4.134e-01	4.310	1.63e-05 ***
CNDN_PCT	8.086e-01	1.233e+00	0.656	0.51184
DEPT1[T.AG]	1.407e+00	1.529e+00	0.920	0.35764
DEPT1[T.AP]	-2.829e-01	8.255e-01	-0.343	0.73178
DEPT1[T.AR]	5.894e-01	1.028e+00	0.573	0.56645
DEPT1[T.CO]	3.149e-01	1.094e+00	0.288	0.77352
DEPT1[T.DE]	4.132e+00	1.695e+00	2.438	0.01479 *
DEPT1[T.ED]	-1.405e-01	1.150e+00	-0.122	0.90272
DEPT1[T.FN]	2.435e+00	1.727e+00	1.410	0.15853
DEPT1[T.FR]	4.229e-01	2.256e+00	0.187	0.85130
DEPT1[T.GR]	1.264e+00	1.059e+00	1.194	0.23263
DEPT1[T.HK]	-1.286e+01	5.415e+02	-0.024	0.98106
DEPT1[T.LA]	1.620e+01	8.111e+02	0.020	0.98407
DEPT1[T.LI]	2.047e+00	1.401e+00	1.462	0.14387
DEPT1[T.ME]	1.749e+01	1.455e+03	0.012	0.99041
DEPT1[T.MU]	3.618e-01	1.337e+00	0.271	0.78660
DEPT1[T.ND]	5.301e+00	1.633e+00	3.246	0.00117 **
DEPT1[T.NU]	-3.497e-01	1.044e+00	-0.335	0.73778
DEPT1[T.PE]	-4.918e-01	1.431e+00	-0.344	0.73117
DEPT1[T.PH]	1.382e+00	1.794e+00	0.770	0.44107
DEPT1[T.PO]	1.799e+01	4.550e+02	0.040	0.96847
DEPT1[T.RH]	1.653e+01	1.455e+03	0.011	0.99094
DEPT1[T.SC]	2.162e-01	1.122e+00	0.193	0.84721
DEPT1[T.SO]	2.328e-01	1.465e+00	0.159	0.87375
DWEL_VAL	1.559e-06	3.960e-07	3.936	8.29e-05 ***
ENG_PCT	-1.073e+00	6.864e-01	-1.563	0.11805
FACSTAFF[T.Y]	1.652e+00	2.539e-01	6.504	7.83e-11 ***
FACULTY1[T.AP]	1.519e+00	1.434e+00	1.059	0.28955
FACULTY1[T.AR]	1.386e+00	1.329e+00	1.043	0.29684
FACULTY1[T.CO]	1.963e+00	1.392e+00	1.410	0.15860
FACULTY1[T.DE]	NA	NA	NA	NA
FACULTY1[T.ED]	2.086e+00	1.439e+00	1.449	0.14723
FACULTY1[T.FR]	1.460e+00	2.452e+00	0.595	0.55151
FACULTY1[T.GR]	2.014e+00	1.292e+00	1.558	0.11912
FACULTY1[T.LA]	-1.216e+01	8.111e+02	-0.015	0.98804

Table 9.5: (Continued)

FACULTY1[T.ME]	-1.436e+01	1.455e+03	-0.010	0.99213
FACULTY1[T.ND]	NA	NA	NA	NA
FACULTY1[T.PH]	NA	NA	NA	NA
FACULTY1[T.PO]	NA	NA	NA	NA
FACULTY1[T.SC]	1.375e+00	1.471e+00	0.934	0.35026
HH_1PER	4.089e+00	2.126e+00	1.923	0.05447 .
HH_2PER	3.218e+00	2.094e+00	1.537	0.12430
HH_3PER	4.980e+00	2.422e+00	2.056	0.03981 *
HH_45PER	4.995e-01	2.354e+00	0.212	0.83195
MOV_DWEL	-1.512e+00	2.160e+00	-0.700	0.48382
OTHERACT[T.Y]	7.143e-02	2.235e-01	0.320	0.74923
OWN_PCT	1.734e-01	4.358e-01	0.398	0.69068
PARENT[T.Y]	5.840e-02	2.011e-01	0.290	0.77151
PROV[T.BC]	-2.333e-01	3.705e-01	-0.630	0.52898
PROV[T.MB]	-5.824e-01	8.783e-01	-0.663	0.50730
PROV[T.NB]	-1.474e+01	7.618e+02	-0.019	0.98456
PROV[T.NF]	1.184e+00	1.379e+00	0.859	0.39061
PROV[T.NS]	1.588e-01	1.539e+00	0.103	0.91783
PROV[T.NT]	-1.353e+01	1.455e+03	-0.009	0.99258
PROV[T.ON]	-7.819e-02	4.154e-01	-0.188	0.85070
PROV[T.QC]	-2.410e+00	9.235e-01	-2.609	0.00907 **
PROV[T.SK]	-1.020e+00	8.007e-01	-1.273	0.20284
SD_INC	-1.764e-06	7.805e-07	-2.259	0.02385 *
SEX[T.M]	4.267e-01	1.311e-01	3.255	0.00113 **
SPOUSE[T.Y]	1.977e-01	1.303e-01	1.517	0.12920
YRFDGR	4.079e-02	2.102e-02	1.941	0.05231 .
YRLDGR	2.173e-02	2.128e-02	1.021	0.30714

Signif. codes: 0 '***' 0.001 '**' 0.01 '*' 0.05 '.' 0.1 ' ' 1

(Dispersion parameter for binomial family taken to be 1)

 Null deviance: 2614.5 on 1885 degrees of freedom
Residual deviance: 1795.9 on 1825 degrees of freedom
AIC: 1917.9

Number of Fisher Scoring iterations: 14

```
> 1 - (WesLogis$deviance/WesLogis$null.deviance) # McFadden R2
[1] 0.3131091
```

Figure 9.9: The Stepwise Variable Selection Dialog

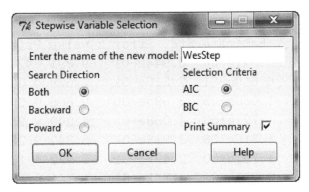

other way to amuse yourself for a few minutes (depending on the speed of your computer) as R chews up CPU cycles comparing different models. When R is done, the model summary shown in Table 9.6 will be printed.

The results indicate that the AIC has been reduced from 1917.9 for the maximal model to 1886.5 for the final model that emerges from the stepwise variable selection process. The McFadden R^2 is just under 0.30, still a high value. The model is also a lot easier to interpret, and is fairly consistent with our original mental model. The somewhat surprising variables are the geodemographic variables HH_1PER, HH_2PER, and HH_3PER. Combined they indicate that a Wesbrook-level donor is likely to live in areas where the size of the typical household is small. Our interpretation is that this is getting at the age of the donor, which is consistent with the strong effect of the years since the donor received his or her first UBC degree. In general, the results point to older, wealthy males who were involved in athletics at UBC, were in the UBC Dentistry, Law, or Medicine Faculties (or did not go to UBC at all), and have children or spouses who also went to UBC as the most likely Wesbrook donors.

Examine the cumulative response chart for the validation sample for the WesLogis and WesStep models. The target level for the lift chart is Y and the true response rate is 0.01 (9.10). The figure indicates that the reduction in variables from 23 to 13 has had no ill effect on the predictive capability of the model.

Step 4: Nonlinear Relationships

The main purpose of the neural network analysis is to determine if there are important nonlinear patterns between the target variable and the set of predictor variables we have selected based on the earlier regression and decision tree analyses. Since our objective is to assess the extent of nonlinear patterns,

Table 9.6: The Logistic Regression Results after Stepwise Variable Selection

```
Coefficients:
                 Estimate Std. Error z value Pr(z)
(Intercept)     -6.521e+00  6.944e-01  -9.391 <2e-16 ***
ATHLTCS[T.Y]     6.610e-01  2.515e-01   2.629 0.008569 **
AVE_INC          1.060e-05  3.696e-06   2.868 0.004128 **
CHILD[T.Y]       1.721e+00  4.058e-01   4.240 2.23e-05 ***
DWEL_VAL         1.716e-06  3.213e-07   5.340 9.27e-08 ***
FACSTAFF[T.Y]    1.731e+00  2.497e-01   6.932 4.16e-12 ***
FACULTY1[T.AP]  -6.398e-01  4.772e-01  -1.341 0.180026
FACULTY1[T.AR]   1.181e-01  4.426e-01   0.267 0.789512
FACULTY1[T.CO]   3.758e-01  4.723e-01   0.796 0.426185
FACULTY1[T.DE]   2.175e+00  6.220e-01   3.496 0.000472 ***
FACULTY1[T.ED]  -9.490e-02  4.959e-01  -0.191 0.848224
FACULTY1[T.FR]  -5.168e-02  6.472e-01  -0.080 0.936346
FACULTY1[T.GR]   5.769e-01  4.401e-01   1.311 0.189916
FACULTY1[T.LA]   2.092e+00  4.754e-01   4.400 1.08e-05 ***
FACULTY1[T.ME]   1.071e+00  4.804e-01   2.230 0.025756 *
FACULTY1[T.ND]   3.335e+00  4.611e-01   7.232 4.76e-13 ***
FACULTY1[T.PH]  -5.532e-01  8.634e-01  -0.641 0.521742
FACULTY1[T.PO]   1.469e+01  2.651e+02   0.055 0.955802
FACULTY1[T.SC]  -3.044e-01  6.444e-01  -0.472 0.636655
HH_1PER          3.089e+00  5.866e-01   5.266 1.39e-07 ***
HH_2PER          2.139e+00  7.552e-01   2.833 0.004614 **
HH_3PER          4.179e+00  1.578e+00   2.648 0.008107 **
SD_INC          -1.552e-06  7.252e-07  -2.140 0.032331 *
SEX[T.M]         4.058e-01  1.240e-01   3.273 0.001064 **
SPOUSE[T.Y]      1.960e-01  1.258e-01   1.558 0.119197
YRFDGR           6.169e-02  5.474e-03  11.268  < 2e-16 ***
---
Signif. codes:  0 '***' 0.001 '**' 0.01 '*' 0.05 '.' 0.1 ' ' 1

(Dispersion parameter for binomial family taken to be 1)

Null deviance: 2614.5  on 1885  degrees of freedom
Residual deviance: 1834.5  on 1860  degrees of freedom
AIC: 1886.5

Number of Fisher Scoring iterations: 13
> 1 - (WesStep$deviance/WesStep$null.deviance) # McFadden R^2
[1] 0.298345
```

Figure 9.10: The WesLogis and WesStep Cumulative Captured Response Chart

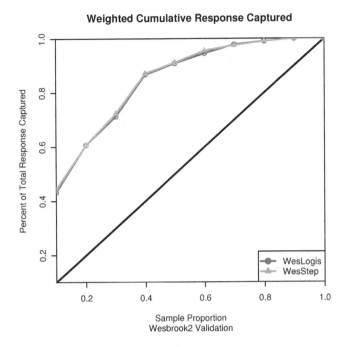

it is appropriate to use a network architecture that allows for a reasonably high level of nonlinearity, so we will use three nodes in the hidden layer. We will also only use the variables in the model obtained through stepwise variable selection: (1) AHLTCS; (2) AVE_INC; (3) CHILD; (4) DWEL_VAL; (5) FACSTAFF; (6) FACULTY1; (7) HH_1PER; (8) HH_2PER; (9) HH_3PER; (10) SD_INC; (11) SEX; (12) SPOUSE; and (13) YRFDGR (incidentally, this also includes the complete set of variables in our initial tree model).

You can try other variables as well, and probably would if time permitted. However, it is unlikely that you would find much improvement, and our goal is to find a "reasonably good" model. Note that the more hidden nodes you have, and the more variables, the longer it will take the neural network model to run. Moreover, and more important, the more likely the model will converge to a local optimum as opposed to a global optimum; in other words, the more likely the neural network model will fail for numerical reasons.

Select Model → Machine Learning → Neural network model... and in the dialog box name the model WesNnet, and make the subset expression Sample == "Estimation," next select WESBROOK as the target, and the above variables selected in the stepwise regression as predictors. Set 3 hidden layer units, and the decay rate to 0.10 (Figure 9.11).

Figure 9.11: Estimating a Neural Network Model

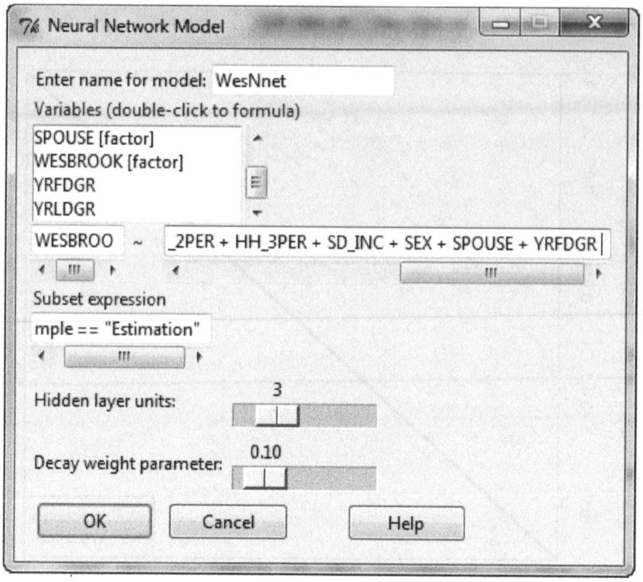

Select **Assess → Graphs → Lift chart...** and choose the three models WesNnet, WesTree2, and WesStep. Select Wesbrook2 as the data set, and make sure that Summary == "Validation" is the subset expression. Set the remaining parameters as before. Click **OK**, and be patient, as it will take several seconds to plot the chart (Figure 9.12). An examination of this particular lift chart indicates that the neural network model does slightly better in the validation sample than the logistic regression model, which means there are some nonlinear patterns in the data that we could consider capturing.

An annoying technical detail is that you may not get exactly this result. In estimating the model, the neural network algorithm can get stuck in a local optimum, which is not the global optimum. In other words, it has not found the best solution. If you see a lift chart that is quite a bit below the others, the algorithm has been trapped in a local optimum. You may be able to find the global optimum, or at least close, by rerunning the model with a higher decay weight, say 0.2.

Step 5: Find the Source of Nonlinearities

If the neural network gives better lift, and we cared about which variables were having what type of effect (which we should, so that we can check that the model doesn't go greatly against managerial experience), we would continue with the methods we learned in Chapters 4 and 5). Specifically we would use the various graphical exploratory tools to examine relationships between WESBROOK and the predictor variables to try and find variables that might

Figure 9.12: The WesLogis and WesNnet Cumulative Response Chart

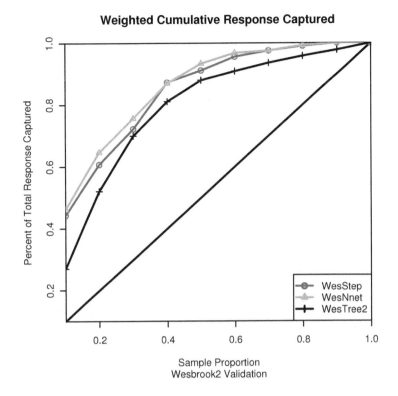

be transformed and entered into the logistic regression. We would work to find a logistic regression model that had the same validation lift as our best neural network, and which could then be interpreted and discussed with management. We will not repeat that analysis here, although the improvement in Figure 9.12 is just enough that it could be worthwhile if there were large numbers of customers.

Scoring the Database

The lift chart helps to decide how many to contact, but does not identify exactly whom they are. To really put it all together, you need to know whom to contact. The prediction and sorting that is done internally to create the lift chart needs to be added to the data set so that an identification variable of the most likely customers (donors in this case) can be matched up with their contact information, often located in another database to protect customers' privacy. Adding this additional information is referred to as *scoring* a database. The chosen predictive model can be used to score any database that has the same variables as used in the model, either the original database, or a brand new database where we do not have the target variable (and hence want to

Figure 9.13: Score a Database Dialog

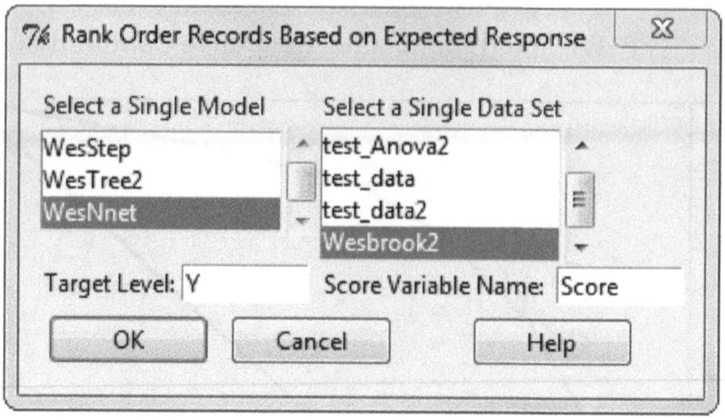

predict it). In this example we will be scoring the Wesbrook2 database we have been working with based on the neural network model we developed. To score a new database we would need to address missing value issues in that database in a way that is similar to what we did with the analysis database. In addition, if we had used variable transformations on some of the predictor variables in the analysis database, those same transformations would need to be performed on the new database.

To score the Wesbrook2 database, use the option **Score → Rank Order** ..., which brings up the dialog box shown in Figure 9.13. Select the WesNnet model in the model selector, and Westbrook2 in the data set selector. The remaining default values should be appropriate. Once you have done this press **OK**. This will cause the new variable `Score` to be appended to the Wesbrook2 database. **View the data set to check to see if this was properly done**. The Score variable provides a rank order number that runs from 1 (best prospect) to 2698 (worst prospect) for each record in the Wesbrook2 database.

You may also wish to have access to the predicted probabilities, rather than merely the ordinal ranking score. To add those to the data set, select **Score→ Sample Estimated Probabilites . . .**, which will bring up a similar dialog box, and add the estimated response probabilities using the selected model and predictors from the selected data. If the sample has been oversampled and you wish to correct for that, select **Score→Adjusted Estimated Probabilities**, and put the true response rate in the dialog (as in the lift chart dialog). The estimated response rates in the original data will now be added to the data set.

Can a Better Model for the Wesbrook Data be Developed?

This model for the Wesbrook data is actually likely a very good "80%" model. We can think of three possible directions to follow to improve this model. First, we would look at ways of sorting out the differences between `DEPT1`, `FACULTY1`, and `MAJOR1`. The second issue that can probably be fruitfully explored is a deeper investigation of the behavior of those people who donate at the Wesbrook level even though they did not graduate from UBC. It may make sense to divide these individuals into different subgroups based on whether other members of their family attended the university, or whether they are current UBC faculty or staff. Finally, our hunch is that income likely has a nonlinear effect. Our suspicion is driven by the fact that both AVE_INC and SD_INC are kept in the stepwise logistic regression model, but have different signs in that model. Given the high level of correlation between these variables, the observed pattern is likely due the model mimicking a nonlinear income effect. These potential nonlinear effects can be addressed. However, cleanly interpreting the meaning of a nonlinear transformation for a geo-demographic variable (which is an average over an area) is actually something of a challenge.

Part III

Grouping Methods

Chapter 10

Ward's Method of Cluster Analysis and Principal Components

10.1 Summarizing Data Sets

Data mining, or any form of statistical analysis for that matter, can be viewed as a set of methods to summarize large amounts of data so that we can usefully interpret the data. Collapsing the data by simply grouping it is common and useful. Cluster analysis and principal components are two broad classes of methods for grouping data. Since we will be using both in the tutorials that follow, a brief explanation of the difference is warranted.

Consider a typical data table that has one row for each individual (a row is also called a "case" or "record"). Across the top, we have the names of the variables (or "fields") which describe the individuals, and down the side we have an individual's identifier (a name or number).

- In principal components analysis (and in a closely related set of models known as "factor analysis"), we search for variables with similar patterns across all individuals. A simple example is the relationship between measures on the intention to purchase a product, and the degree of liking for that product. Most likely, both would be high or both low for any one individual. For this (or more complex cases) we can combine several variables into a single variable, allowing us to collapse the data table horizontally. We will end up with fewer variables, now called *components* or *factors*, describing our data.

- In cluster analysis we are looking for individuals who have similar patterns across all variables. Similar individuals can be combined into a group, and be represented by the group average on each variable, allowing us to collapse the data table vertically. We have fewer rows, now called clusters, describing our data.

In this chapter and the next we will introduce two methods of cluster analysis which have complementary advantages and disadvantages. We would also like to have a graphical visualization of the clusters, that is, a plot of the clusters' averages of the variables on which our individuals are measured. Since traditional plots are restricted to two, or at most three, dimensions, we will also need to collapse the variables in order to create the plots. To do this, we will introduce principal component analysis to reduce the number of variables in the data for plotting purposes.

Cluster analysis is used in many fields, and is called by many different names:

- *Numerical taxonomy* by biologists

- *Unsupervised pattern recognition* by computer scientists

- *Clumping* by geographers

- *Partitioning* by graph theorists

- *Seriation* by anthropologists

- *Segmentation* by marketers

The clustering methods will be introduced in this book in the context of segmentation. Once the techniques are mastered, we will move on to practical issues involved in applying these methods to segmentation problems, such as determining which variables, of those available, we should use to form our segments.

10.2   Ward's Method of Cluster Analysis

Clusters of individuals in a data set can be generated in three ways. First, we can specify a set of clusters a priori and then find the "best" partitioning of the data into the specified number of clusters based on some criteria; this type of approach is known as a partitioning method. Second, we can start with the entire group of individuals in the data set as one large cluster, and split it into a number of smaller clusters, which is a method known as divisive hierarchical clustering. Our third option is to start with each individual as a single cluster, and combine them into larger clusters (starting with pairs of individuals), an approach called agglomerative hierarchical clustering. Ward's method (Ward, 1963) uses the third of these different approaches.

Whenever we use any method to summarize data, we increase the ease of interpretability of our data, but we lose detailed information. The Ward's method algorithm selects the individuals that are to be added to a group at each stage based on minimizing the loss of information that occurs when the group is described by the *average* values of the group members' variables, rather than the many values for all of the members. Two individuals who have identical values for the variables being used to cluster on can be combined and represented by their average with no loss of information. The more different the individuals are, the more information is lost when they are represented by their cluster average.

The key features of the Ward's algorithm:

- Only continuous variables can be used in a Ward's method cluster analysis.

- The average value of a cluster is referred to as the cluster center or *centroid*.

- The loss of information is based on distances or on squared distances. In the latter case we use "error sum of squares" or ESS as the measure of loss.

- The ESS is the sum of squares of distances of each individual from his or her cluster center.

- The distance is usually Euclidean. With n variables, the Euclidean distance between two points $x = (x_1, x_2, x_3, \ldots, x_n)$ and $y = (y_1, y_2, y_3, \ldots, y_n)$ is

$$D_{x,y} = \sqrt{(y_1 - x_1)^2 + (y_2 - x_2)^2 + \ldots + (y_n - x_n)^2}.$$

- With m_i individuals $x_{i,j}$ in a cluster i with centroid c_i, the ESS_i for the cluster is

$$ESS_i = \sum_{j=1}^{m_i} D^2_{c_i, x_{i,j}}.$$

- The ESS for a particular clustering scheme consisting of s clusters is

$$ESS = \sum_{i=1}^{s} ESS_i.$$

- The algorithm starts with each individual as a cluster of one.

- The two individuals with the smallest $ESS_{(2)}$ are joined to form the first cluster, where the subscript "(2)" indicates a two-member cluster.

Figure 10.1: Customer Locations along the Rating Scale

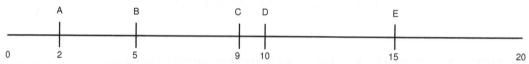

- The algorithm then either joins two other individuals who have the smallest $ESS_{(2)}$, forming two segments of two members; or joins one individual with the first segment of two members, forming the first segment with the smallest $ESS_{(3)}$.

- Repeat, at each stage selecting the cluster scheme with smallest total ESS.

10.2.1   A Single Variable Example

A numeric example will help clarify the process. To keep things simple, the example has only one variable, so that the distances are simply differences in the variable values.

Suppose we survey five customers and ask them to rate their intention to purchase a new car within the next year. Our customers are unimaginatively named A, B, C, D, and E, and provide ratings of 2, 5, 9, 10, and 15, respectively, on a 20-point scale. To help visualize the relationship between respondents, Figure 10.1 shows the customers' locations along the rating scale.

To cluster these customer ratings, the algorithm proceeds as follows:

First Stage: Five clusters of one, with ESS = 0, and no loss of information since there is no clustering!

Second Stage: Search for the two that can be combined to give the smallest $ESS_{(2)}$.

That is C and D — obviously — which becomes one cluster, with centroid $(9+10)/2 = 9.5$.

This is a four "cluster" solution, with clusters A, B, CD, and E. The ESS for the new four-cluster solution is $0 + 0 + [(9 - 9.5)^2 + (10 - 9.5)^2] + 0 = \mathbf{0.5}$.

Third Stage: Search all possible three cluster schemes, *which include the CD cluster*, to find the one that gives the smallest ESS;

1) ESS for the solution consisting of AB, CD, and E is

AB: $(2 - 3.5)^2 + (5 - 3.5)^2 = 4.5$; CD $= 0.5$; and E $= 0$

ESS $= 4.5 + 0.5 + 0 = $ **5.0**

2) ESS for the solution consisting of A, B, and CDE is

A and B: 0

CDE centroid $= (9 + 10 + 15)/3 = 34/3 = 11.33$

$(9 - 11.33)^2 + (10 - 11.33)^2 + (15 - 11.33)^2 = 20.66$

ESS $= 0 + 0 + $ **20.66**

Similarly, the ESS for the remaining possible three-cluster solutions are

3) AE, CD, and B; 4) A, BE, and CD; 5) A, BCD, and E; 6) ACD, B, and E;

    ESS $= $ **85.0**   ESS $= $ **50.5**   ESS $= $ **14**   ESS $= $ **38.0**

The minimum ESS for any of the three-cluster solutions is **5.0**, and consists of the clusters AB, CD, and E.

Fourth Stage: Compare the ESS of all possible two-cluster schemes, *starting from the best three-cluster solution.*

1) ESS for the solution consisting of ABCD and E

Centroid of ABCD: $(2 + 5 + 9 + 10)/4 = 6.5$

$(2 - 6.5)^2 + (5 - 6.5)^2 + (9 - 6.5)^2 + (10 - 6.5)^2 = 41$

ESS $= 41 + 0 = 41$

2) ESS for the solution consisting of ABE and CD: **ESS $= $ 93.17**

3) ESS for the solution consisting of AB and CDE: **ESS $= $ 25.18**

The minimum ESS for a two-cluster solution is **25.18**, and consists of the clusters AB and CDE.

Fifth Stage: only one possible one-cluster solution, ABCDE:

Centroid $= (2 + 5 + 9 + 10 + 15)/5 = 41/5 = $8.2

ESS $= $ **98.8**

A few points to note:

- At each stage, we work with the clusters from the previous stage — we never take a previous cluster apart. This creates a hierarchical structure of clusters, and Ward's method is thus a *hierarchical* method. The hierarchy of clusters can be represented as a dendrogram (in simple terms, a tree diagram, see Figure 10.2), which shows the individuals in each cluster and the ESS at each stage. The individuals are labeled on the x-axis, but there is no attempt to specify their distance along the x-axis. We could do that in our one-dimensional example, but cannot in multi-dimensional problems. All that is indicated on the x-axis is that nearby individuals are similar. The degree of similarity is indicated by the information loss in clustering (the ESS), which is plotted on the y-axis.

- The analyst uses the dendrogram to help choose the number of clusters to ultimately work with, by balancing the increase in interpretability and applicability gained using fewer clusters, with the greater information retained by using more clusters. This is a judgment call that depends on the application.

- As the number of records in the database increases, it eventually becomes impossible to identify the individuals on the x-axis — it becomes too crowded. More critically, as the number of individuals increases, the computation time rapidly increases as you might guess from this simple example. This renders Ward's method unsuitable for large databases.

- Measures other than distance-squared can be used. One common alternative is the total distance. In that case, instead of the Error Sum of Squared Distances, the information loss is measured by the sum of the Euclidean distances.

10.2.2   Extension to Two or More Variables

Clustering on two variables is illustrated in the stylized example in Figure 10.3, which shows the results of a survey that includes customers' importance ratings of strength and water resistance of carpet fibers. Each pair of ratings for a respondent is plotted as a single point in a two-dimensional plane. The figure shows a four-cluster solution, with the four cluster centers (shown as black dots). The ESS is calculated in a similar fashion to the one-dimensional case. The algorithm proceeds similarly, starting by grouping the closest two individuals into a cluster, continuing through the illustrated four-cluster solution, and ending with a single cluster.

Working with more variables is algebraically straightforward — the distance

Figure 10.2: A Dendrogram Summarizing a Ward's Method Cluster Solution

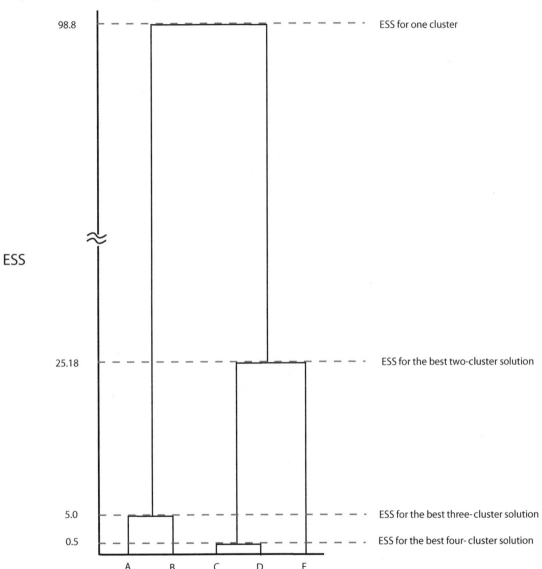

Figure 10.3: A Two Variable Cluster Solution Presented as a Two-Dimensional Plot

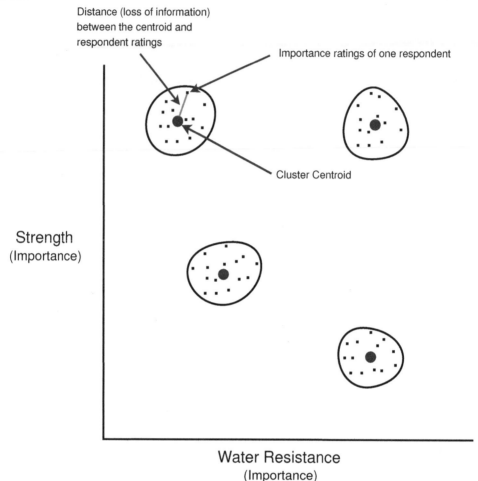

measure is just adjusted appropriately. We cannot plot the results in a straightforward fashion. However, the dendrogram can always be plotted, since it does not show individual variables.

10.3 Principal Components

Principal components (Mardia et al., 1979) summarizes data across variables by transforming the original set of variables into a smaller set that accounts for most of the variance (i.e., information) in the data. The new variables are called factors, or principal components. It is a useful data exploration tool in its own right, but our use here will be to help us with data visualization of the cluster analysis results when there are many variables. With more than two or three variables, displaying the results in two- or three-dimensional plots directly is impossible. If we can capture most of the variation in two or three principal components, we can plot our results using them as axes.

The principal components are extracted sequentially, with the first principal component accounting for the most variance in the data. If the data consist of p variables, $x_1, x_2, \ldots, x_p$, the first principal component, $PC_{(1)}$, is given by a linear combination of the variables

$$PC_{(1)} = w_{(1)1}x_1 + w_{(1)2}x_2 + \ldots + w_{(1)p}x_p,$$

where the weights $w_{(1)j}$ are chosen to maximize the variance of $PC_{(1)}$ across all of the individuals in the data. Intuitively, $PC_{(1)}$ is a vector that points in the direction in which the data are most "spread out." If this looks vaguely like a linear regression model, it is because in both cases the right-hand side is a linear combination of the variables. However, in linear regression, we choose the weights, or coefficients, to minimize the squared difference between the right-hand side and an observed target variable (typically labeled "y") placed on the left-hand side. The second principal component, $PC_{(2)}$, is set up similarly, but with the additional constraint that it must be orthogonal (or uncorrelated) to $PC_{(1)}$. The third must be orthogonal to both of the previous two. We can continue on, finding as many principal components as there are original variables. Each one will capture less variance than the previous one, and together they can be used to form a new orthogonal coordinate system.

Some points to note:

- Since the purpose is to reduce the dimensionality of the data, we will almost always use far fewer components in our interpretation than there were original variables. The number of principal components to retain depends on the intended purpose, and is a judgment call. It depends on how much information we are willing to lose for the sake of parsimony and interpretability. The information loss is assessed by the *proportion of the variance in the original data that is captured by successive principal components*. In our case, where our objective is plotting, we will be constrained to retaining only the first two or three components.

- To help interpret the results, we need to attach some meaning to each principal component used. This is facilitated by calculating the correlations, called component *loadings*, between the original variables and the principal components. This allows us to see which variables contribute most to which components. If we use the components as a coordinate

Table 10.1: Attribute Ratings for Seven Potential Employers

Employer	Family Life	Location	Challenge	Financial	Learning
Accenture	1	3	8	8	7
Hershey's	9	4	4	4	5
GE Brazil	5	6	6	5	3
Startup	2	8	9	2	8
Family Business	8	9	4	3	4
Safeway	7	8	3	4	5
Statistics Canada	5	3	5	3	6

      system, we can plot the loadings for each variable in that coordinate system, which also helps with interpreting the principal components. We illustrate the process of doing this in the example below.

- Each individual in the data set can be represented in this new coordinate system simply by using the transformation from each individual's original coordinates to the principal component coordinates, or *component scores*. Aside from being a handy initial data reduction technique when there are many variables, this also allows us to represent our data graphically in the new reduced coordinate system.

The following simple example illustrates the interpretation of principal component analysis graphically. Suppose a recent star business graduate with strong analytics training receives seven job offers, and to help her decide which to select, she identifies five important characteristics, and rates each offer on those characteristics, using a scale that goes from 1 to 9, with the values of these ratings shown in Table 10.1.

Do you see any patterns in these data? In particular, do some of the characteristics generate similar, or possibly opposite, patterns of ratings across the potential employers? **Take a minute to study the numbers with this question in mind.**

We can use graphical visualization tools to help out. The first thing we could do is plot the variables individually across the different job offers. For example, we could compare the challenge, learning opportunities, and quality of family life of the jobs, as shown in Figure 10.4. **Again, take a moment to inspect these plots — you should be able to answer the questions in the figure.**

It looks like challenge and learning opportunities have similar patterns, and both are the opposite of quality of family life. Another way to show this is with

Figure 10.4: Comparing Offers on Different Attributes

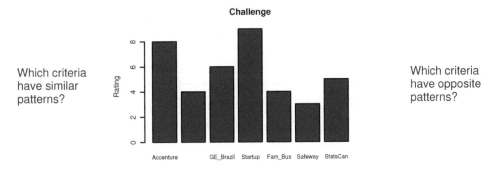

Which criteria
have similar
patterns?

Which criteria
have opposite
patterns?

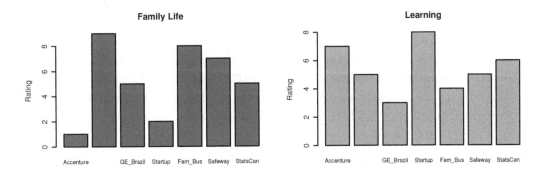

Figure 10.5: Family Life vs. Challenge for Different Jobs

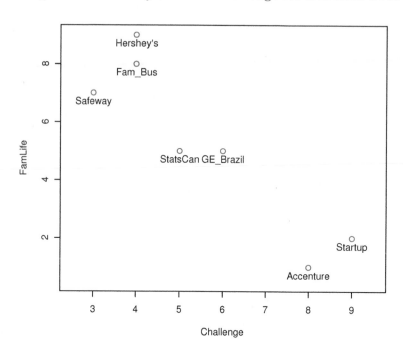

a bivariate scatterplot, as shown in Figure 10.5. Here we see that as challenge increases across jobs, the quality of family life tends to decrease. We could also express this as a negative correlation between the two job characteristics.

Principal components analysis takes all of the variables and cases at once, rather than only two at a time as in the scatterplot, and allows us to plot the results to give a summary of all of the relationships. In Figure 10.6, we show a two-dimensional plot with the first and second principal components as the axes, which is often called a *bi-plot*.

Within this coordinate system, we can use the results of the analysis to plot both the original variables and the job offers.

Interpretation points:

- Arrows represent the original variables. Arrows that are nearly parallel and in the same direction indicate variables which are highly positively correlated (Learning and Challenge); arrows pointing in opposite directions are highly negatively correlated (Challenge and FamLife); and arrows that are perpendicular are uncorrelated, or *orthogonal* (Location and Learning).

- Increasing values of PC1 indicate mainly improved family life and re-duced learning and challenge opportunities, and to a lesser extent, better

Figure 10.6: Principal Components of the Employer Ratings Data

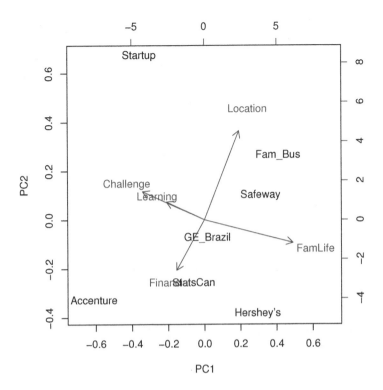

location and worse financial rewards. Increasing values of PC2 indicate more desirable locations and lower financial rewards, and to a lesser extent, increased learning opportunities.

- In the tutorial, we will keep track of the variance captured by the principal components as a measure of information retained (or lost) as we reduce the dimensionality of the data. You can also get some sense of how much information is retained (or lost) by comparing the locations of the firms in the plot with the original data. *Family Business* is plotted the farthest in the direction of the *Location* vector (up and right), and also scores the highest on *Location* in the original data. *Family Business, Safeway*, and *Hershey's* are furthest, and nearly equally far, in the direction of the *FamLife* vector (right); they are also the three highest rated on that dimension in the original data. *Accenture* is plotted furthest in the direction of *Financial* rewards, and is also plotted furthest in the *opposite* direction of *Location* as well as rated high and low, respectively, on these two characteristics in the original data. This is all encouraging, in that we have reduced the data from five dimensions to two dimensions and retained these ordering relationships. However, we simply cannot retain all of the original information. **Find some order-**

ing relationships in the original data that have been lost in the two-dimensional principal component plot.

10.4   Ward's Method Tutorial

This tutorial will give you hands-on experience in using R Commander's hierarchical cluster analysis for market segmentation and principal components analysis to reduce the number of variables to facilitate interpretation of the cluster solution. For those who are familiar with factor analysis, "principal components" are very similar to "factors." In fact, one class of factor analysis methods begins with a principal components solution, which is then adjusted. You will also see how to create two- and three-dimensional plots of a clustering solution using principal components, which is handy when many variables are used in the cluster analysis.

The data used in this lab are from a study of the relative importance that different stakeholder groups at a particular U.S. university place on seven different measures of intercollegiate athletic program performance. We will use cluster analysis to segment the respondents on seven importance weights, which themselves are the output of a conjoint analysis, a commonly used market research technique. The source of the data is only relevant because it makes the data unusually well behaved. Normalizing data is desirable prior to cluster analysis when the variables have very different numerical ranges. However, the conjoint analysis based relative importance weights are constructed so that they are all between zero and one, and sum to one for each respondent. Because of these properties, there is no need to normalize these measures before clustering. The clustering of relative importance weights derived from conjoint analysis is also a common marketing research practice. It allows for the construction of a limited number of benefit segments and thus helps in the development of a manageable number of products or campaigns to address the different segments.

1.

Use the pull-down menu option **Data → Get from → R package → Read data from an attached package...** to load the '**Athletic**' data set from the BCA package into R. As always, the first task is to look at the documentation for this data set (which is named Athletic) using **Data → Clean → Help on active data set (if available)** to get a better understanding of the variables used to measure perceptions of intercollegiate athletic program success. Note

Figure 10.7: The Hierarchical Clustering Dialog Box

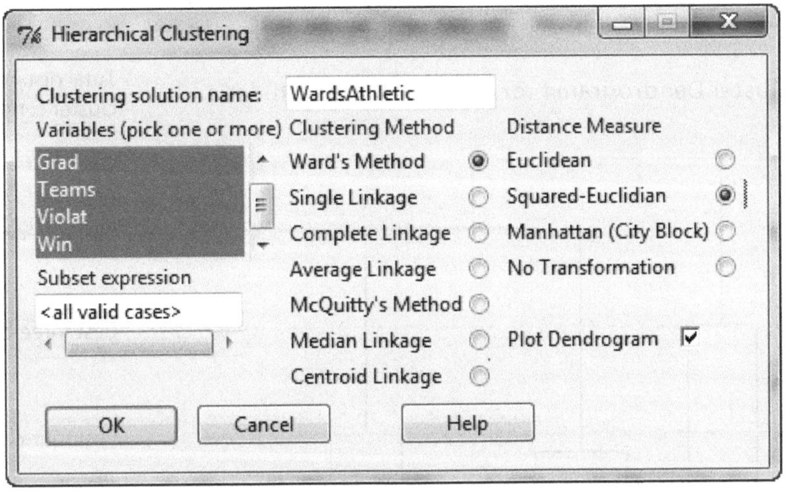

that this particular data set, unusually, contains no categorical variables. The reason for this is that Ward's method of cluster analysis (and the K-Centroids cluster analysis method you will see in the next chapter) can only be appropriately used with continuous variables. **Click on View data set** button to check the variables. By scrolling to the bottom of the window, you can see that the last individual is number 184. However, 26 respondents have been removed because of incomplete responses and missing values, so that there are only 168 individuals. The data have been thoroughly cleaned.

2.

To create the Ward's method clustering solution use the pull-down menu option **Group → Records → Hierarchical cluster analysis. . .**, which will bring up the dialog box shown in Figure 10.7.

3.

Enter `WardsAthletic` in the "Clustering solution name:" field. **Select all the variables** in the data set in the "Variables (pick one or more)" option. You can quickly select all of the variables by clicking and dragging down. Both the default clustering method (Ward's method) and default distance measure (Euclidean) are selected by default. The original version of Ward's method uses squared Euclidean distances, so click on the **Squared-Euclidian** radio button. In addition, the "Plot Dendrogram" option is already checked. **Press the OK button**, and the dendrogram shown in Figure 10.8 will appear in an R graphics window. The graphics window may be behind existing windows on your desktop, so you might have to bring it to the front.

Figure 10.8: The Annotated Ward's Method Dendrogram for the Athletic Data Set

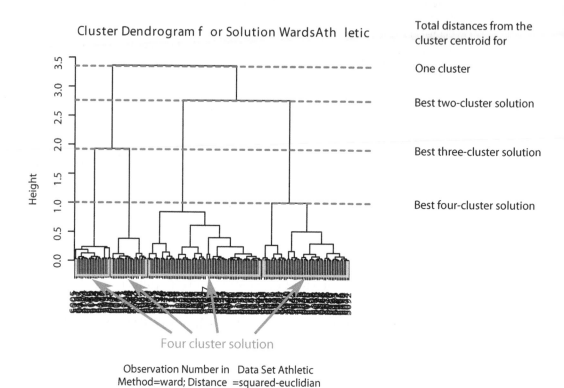

Four cluster solution

Observation Number in Data Set Athletic
Method=ward; Distance =squared-euclidian

The dendrogram shows groupings of individuals with similar importance ratings. At the bottom are 168 "segments of one" and at each stage the algorithm selects the most similar previous two groups to combine. At the top there is a single group of 168 individuals.

To interpret the diagram, start at the top and work down. As the number of clusters increases, they become smaller, of course, and more individuals are going to be closer to their cluster centroids, hence the total distances to cluster centroids decreases. This total distance is a measure of how diverse the individuals are in the clusters. A good rule is to look for a level where splitting a cluster into two gives a large decrease in the diversity, as measured by the total distances. In Figure 10.8, there is a large decrease between three and four clusters, and choosing a four-cluster solution here is reasonable.

4.

At this point inspect the summary of the four-cluster solution. To do this, select the pull-down menu option **Group → Records → Summarize hierar-**

Figure 10.9: The Hierarchical Cluster Summary Dialog Box

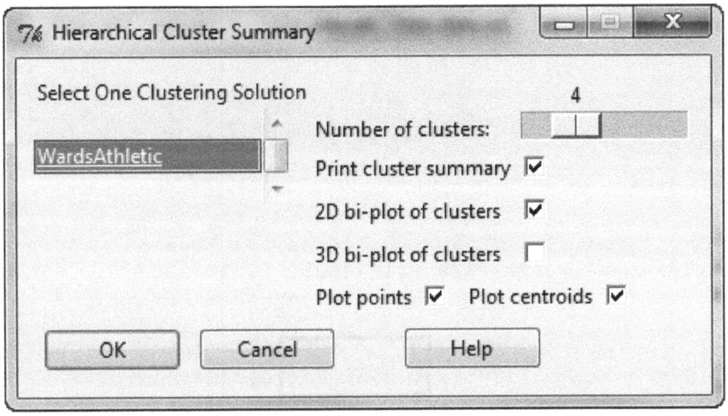

chical clustering..., which will cause the dialog box shown in Figure 10.9 to appear.

5.

In the Hierarchical Cluster Summary dialog box **click on the WardsAtheltic** solution as the "Select One Clustering Solution," use the **slider bar to select 4** as the "Number of clusters." **Leave both the "Print cluster summary" and "2D Bi-plot of clusters"** options checked since this will allow us to examine which aspects of intercollegiate athletic programs provide the basis for the clustering solution. **Check "Plot points" and "Plot centroids"** for now. **Press OK** and the cluster summary shown in Figure 10.10 will appear in R Commander's output window. The bi-plot will appear in the R graphics window and we will save it to examine later. For Windows users, to keep all of the graphics output available for the rest of this tutorial, **click on the History menu in the Graphics window and select Recording**.

6.

The first row gives the number of individuals assigned to each cluster (53, 70, 23, and 22; summing to 168). The clusters are identified by their index number from 1 to 4 in Figure 10.10. Each cluster is summarized by the average of the seven importance values for the individuals within a cluster. As a first cut at interpreting the clusters, we can note what the respondents in each cluster feel is the most important criterion for an intercollegiate athletic program to be called successful. For example, cluster 2 believes that athlete graduation rates are most important. For each cluster, the seven averages together are referred to as the *cluster center* or *centroid* for that cluster.

Figure 10.10: Cluster Centroids for the Four Ward's Method Clusters

```
> summary(as.factor(cutree(WardsAthletic, k = 4))) # Cluster Sizes
 1  2  3  4
53 70 23 22

> by(model.matrix(~-1 + Attnd + Fem + Finan + Grad + Teams + Violat + Win,
+   Athletic), as.factor(cutree(WardsAthletic, k = 4)), mean)
+   # Cluster Centroids
INDICES: 1
      Attnd        Fem      Finan       Grad      Teams     Violat         Win
 0.12117511 0.07572518 0.24178826 0.12148291 0.05932957 0.11360185  0.26689712
------------------------------------------------------------
INDICES: 2
      Attnd        Fem      Finan       Grad      Teams     Violat         Win
 0.08353761 0.08302892 0.17880561 0.29502756 0.06245411 0.12757679  0.16775723
------------------------------------------------------------
INDICES: 3
      Attnd        Fem      Finan       Grad      Teams     Violat         Win
 0.06311553 0.06612890 0.15329216 0.17992638 0.05717968 0.37358539  0.10677195
------------------------------------------------------------
INDICES: 4
      Attnd        Fem      Finan       Grad      Teams     Violat         Win
 0.08102743 0.06058633 0.38121874 0.12462090 0.06466165 0.20351505  0.08436991
```

7.

We can also compare the clusters on each of their means graphically, but first
we need to add a variable to the data set that identifies which cluster each
individual belongs to. This identifier will be useful in later analyses as well. To
do this, use the pull-down menu option **Group → Records →Add hierarchical
clustering to data set...**. In the dialog box select **WardsAthletic** as the
"Select One Clustering Solution," in the "Assigned cluster label:" field enter
Wards4, and set the slider to **4** as the "**Number of clusters:**" (Figure 10.11).
Click OK, and the factor variable **"Wards4"** will be added to the Athletic data
set. Its value will be the cluster index to which each respondent is assigned by
the four-cluster Ward's method solution. Press the **View Data Set** button to

Figure 10.11: Append Cluster Groups to Active Data Set Dialog Box

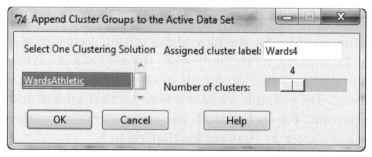

Figure 10.12: The Plot of Means Dialog Box

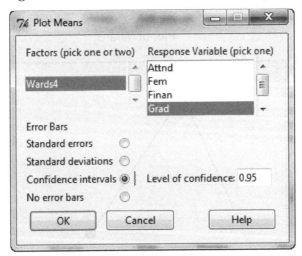

inspect the data and ensure that this is the case. You will likely have to scroll right to see the new variable.

8.

Use **Explore and test** → **Visualize** → **Plot of means. . .** to bring up the dialog in Figure 10.12. **Select** Wards4 as the factor variable for grouping, and Grad as the response variable. **Select Confidence intervals (95%)** to see the variation within the cluster. **Click OK.** The means plot (Figure 10.13) shows that cluster 2 cares most about "Grad" and the error bars indicate that this high level of caring is statistically different from the lower level of caring in other three clusters.

9.

If there were only two variables, say Grad and Win, we could visualize the full four-cluster solution by simply plotting the four cluster centers on a two-dimensional plot that has the two variables as axes. However, with seven variables, plotting the cluster centers for easy visualization is a bit tricky. The way to approach this is to collapse the variables using principal components, and then plot just the first two components, which, if we are lucky, will retain enough information to be useful. This type of plot is known as a "bi-plot." We produced the bi-plot in step 5, but it is now behind the means plot in the *graphics window*. In Windows, on the R *Graphics window* menu, select **History** → **Previous** to return the bi-plot, shown in Figure 10.14. Under Mac OS X and Linux you will need to repeat step 5 of this tutorial.

The bi-plot uses the two most important principal components (based on the

Figure 10.13: Plot of Means of Graduation Rates by Ward's Method Clusters

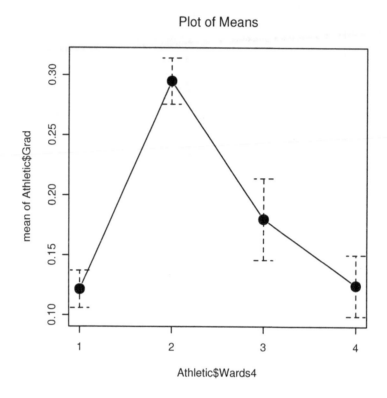

percentage of variance they explain in the data) as the axes, labeled PC1 and PC2, and plots the original seven variables as arrows (vectors) in terms of PC1 and PC2. The location of the arrows in the two dimensional space is interpreted as how much each variable is contributing to each of the two principal components. PC1 consists mainly of WIN and the negative of VIOLAT; PC2 consists mainly of GRAD and the negative of FINAN. The other three variables are not contributing much to the first two principal components. The plot also gives the location of each individual in the space defined by the first two principal components, and cluster assignments for each individual or the cluster centroids, depending on what is checked in the dialog box. In this case, there are few enough individuals that the plot remains interpretable with all of the information contained in it; however if there were many individuals, plotting only the centers of each cluster would be necessary for interpretability.

10.

Interpretation of the bi-plot: The bi-plot in Figure 10.14 does a lot to help us understand differences across clusters in terms of their perceptions of characteristics of a successful intercollegiate athletic program. Four of the bi-plot

Figure 10.14: Bi-Plot of the Ward's Method Solution of the Athletic Data Set

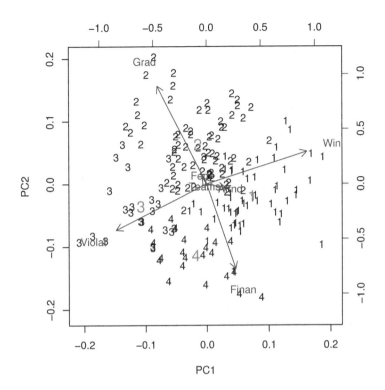

arrows (Finan, Violat, Grad, Win) form a fairly clean coordinate system, and the remaining three variables do not load on either of the first two principal components — this is indicated by the fact that they are very short. The individuals are plotted as the index number for the cluster they belong to, and the centroids are numbers in blue. A nice result in this data is that one cluster is associated with each of these four arrows. This does not always happen! The more southeasterly an individual is in the plot, the more importance they place on financial performance in judging program success, and, as a group, members of cluster 4 are most concerned with financial success. Individuals to the southwest feel that avoiding NCAA violations is important in the success of an intercollegiate athletic program, and members of cluster 3 are most concerned with this issue. Cluster 2, toward the northwest, cares about graduation rates. Finally, as an individual is more to the northeast, the more important win–loss records become in their overall assessment of the success of an intercollegiate athletic program, and is represented by cluster 1. Inspect the clusters' mean importance weights in Figure 10.14 to confirm that they are consistent with the bi-plot interpretation. Note that Figure 10.14 additionally seems to indicate that no group is particularly concerned about either gender

Figure 10.15: Obtaining a Three-Dimensional Bi-Plot

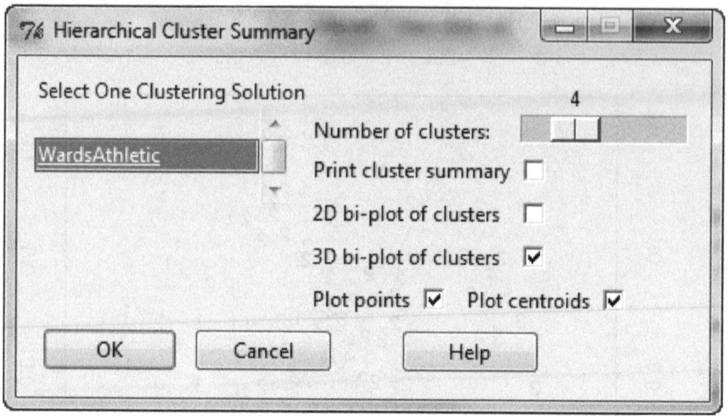

equity (Fem), attendance, or the number of teams. This is consistent with the very short arrows for these variables.

11.

There is one more graphical tool that can help with interpreting the clustering results. While the seven variables appear to collapse nicely into two principal components in this case, often a third component is useful. We will generate a three-dimensional plot of the first *three* components. Again select **Group →Records → Summarize hierarchical clustering. . .** , and as before, **select the WardsAthletic solution, and set the slider to 4.** This time only **check the 3D bi-plot of clusters** box as in Figure 10.15. Click **OK.** A new graphics window labeled "RGL device 1" will appear, but it will be small. You can click and drag the corner of this window to expand it, or click on the upper right corner to expand the window to full screen size. To control viewing the plot (1) hold down the left mouse button and drag the pointer within the window to rotate the plot, and (2) use the scroll wheel to zoom.

12.

First, orient yourself by rotating the plot around to try and make it look somewhat like the two-dimensional bi-plot. You can drag the two graphics windows around on your screen until you can see them both at the same time. The key here is to get PC3 pointing "out of the screen" so that you only have PC1 and PC2 with some length to them, as in Figure 10.16. You can compare this with the two-dimensional bi-plot and see the similarity. To get your depth perception working, move the plot around. You can see that the individuals in each of the four segments are still forming four clouds around the same four variables as in the two-dimensional case.

Figure 10.16: The Three-Dimensional Bi-Plot of the Athletic Data Set

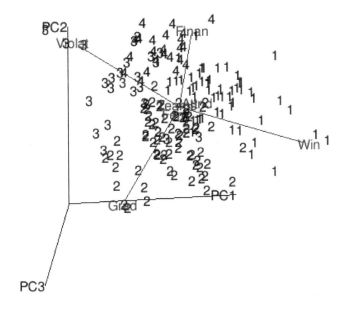

13.

To get some practice in using this tool, rerun the hierarchical cluster analysis selecting only the four important variables. Choose the number of clusters that seems appropriate to you, add it to the data set, and summarize. Note similarities and differences from the four-cluster solution which used all seven importance variables (look at the bi-plots).

14.

You can either continue on to the K-means clustering tutorial, or if you exit, make sure to save your workspace, with a descriptive filename like Wards.RData, before exiting.

Chapter 11

K-Centroids Partitioning Cluster Analysis

The last chapter looked at hierarchical clustering, the first form of clustering analysis developed, which continues to be used in many applications. However, hierarchical clustering methods do not scale well to large databases since they require the calculation of the distance between every record in a database, something that works up to a few thousand records, but becomes problematic beyond that point. Partially to deal with this limitation of hierarchical clustering, as well as for a number of other reasons, researchers in several disciplines have developed other methods of cluster analysis since the pioneering work of MacQueen (1967). The clustering approach most commonly used in applied data mining is known as partitioning cluster analysis, in which the records are divided (partitioned) into the "best" K groups based on some criteria. Nearly all the partitioning cluster analysis methods accomplish their objective by basing cluster membership on the proximity of each record to one of K points (or "centroids") in the data. The objective of these clustering algorithms is to find the location of the centroids that optimizes some criteria with respect to the distance between the centroid of a cluster and the points assigned to that cluster for a pre-specified number of clusters in the data.

In this chapter we describe how K-Centroids clustering algorithms work in general, present three different variants of the algorithm: K-Means, K-Medians, and Neural Gas, and then provide a graphical illustration of how K-Means clustering works. The three K-Centroids variants differ from one another in the distance metric that is used to measure the distance between a cluster centroid and its member points, and in how the centroids themselves are defined. We next take a minor detour and discuss the related issue of the different types of clusters that can exist, and the relationship between these cluster types and the nature of customer segments that are typically found in applied data mining projects. The next issue we explore are methods to determine both the "best" number of clusters and the "best" clustering method for a data set using two different cluster validation measures, the adjusted Rand index (Hubert and Arabie, 1985) to assess cluster solution stability and reproducibility, and the Calinski–Harabasz index (Calinski and Harabasz, 1974) to assess the relative level of within-cluster homogeneity to cross-cluster separation.

11.1   How K-Centroid Clustering Works

In K-Centroid clustering the number of clusters is selected by the analyst, and the algorithm then searches for the "best" way to partition the data into the specified number of clusters. The advantage over hierarchical clustering is that individual cluster members are added and removed from different clusters as the algorithm progresses, leading to improved solutions. In hierarchical clustering, once a cluster is formed the cluster members are fixed; all that can happen is that individual clusters are combined into larger clusters. The disadvantage of K-Centroid clustering is that the analyst has to specify the number of clusters in advance, and it is very difficult to know how many clusters to specify. As a result of these complementary advantages and disadvantages, a common approach is to use hierarchical clustering to select a desirable number of clusters, and then use K-Centroids to refine the membership of that number of clusters. An alternate, and in our opinion better, approach (which we look at in Section 11.3) is to try a number of different K-Centroids cluster solutions (with respect to the number of clusters used), and use one or more cluster validation methods to select possible solutions. Both of these provide the analyst with guidance as to the number of clusters in the data, but they also help indicate whether there is a stable (reproducible) cluster structure in the data. For example, if both a Ward's method solution and a K-Centroids solution are similar in terms of how data points are allocated to clusters, it provides further evidence that there is a well-defined cluster structure in the data. However, if there is no real cluster structure in the data, both methods will still give solutions, but in that case, the solutions can be quite different.

11.1.1   The Basic Algorithm to Find K-Centroids Clusters

The general K-Centroids algorithm is as follows:

1. Choose the number of clusters, K, and assign K starting centroids within the space of the individuals to be clustered.

2. Assign each individual to the nearest centroid, forming K clusters.

3. Calculate a new position for each centroid on some measure of central tendency (such as the mean) for each of the K clusters.

4. Repeat steps 2 and 3 until there are no further changes in the centroid locations.

Some points to note:

- When there are no well-defined clusters in the data, the final clusters depend very heavily on the initial position of the centroids. When there are K well-defined clusters (a topic we discuss in the next section), and K centroids are chosen, the final solution is relatively insensitive to the initial choices.

- To help guard against solutions which are local optima, the algorithm can be run a number of times with different starting centroids, or starting "seeds" and the "best" one, the one with the minimum ESS (error sum of squares), is selected.

- The choice of the methods to measure distance and for calculating centroids can have a strong influence on the nature of the clusters that are uncovered, particularly in cases where the underlying cluster structure is fairly weak.

- While there are a number of computational methods to assist in determining the "right" number of clusters (and the "right" method to use), qualitative assessments of possible clustering solutions need to be carried out by the analyst to investigate whether the clustering solution will be useful for managerial decision making. Typically this is done by "profiling" the clusters in terms of how different the centroids (and the clusters themselves) are from one another, and whether there is a relationship between the clusters and other variables (such as demographic variables) that makes the resulting clusters (segments) useful for target marketing campaigns, developing promotional offers, or developing other types of business programs.

11.1.2 Specific K-Centroid Clustering Algorithms

By far the most commonly used K-Centroid clustering algorithm is K-Means clustering (MacQueen, 1967). In K-Means, the distance measure used is Euclidean distance, and new cluster centroids are calculated by taking the mean of all the variables used to form clusters for the records that are assigned to a cluster. To make this process more concrete, Figure 11.1 provides an illustration of the steps involved in creating a K-Means clustering solution.[1] The (hypothetical) example data, shown in Panel A of the figure, are for a service

---

[1]See http://home.dei.polimi.it/matteucc/Clustering/tutorial_html/AppletKM.html for a nice interactive demonstration applet of the same point. In this applet, the initial cases are randomly distributed (unclustered), but you can move them into clusters, as well as shift the initial centroids, before running the algorithm.

Figure 11.1: Steps in Creating a K-Means Clustering Solution

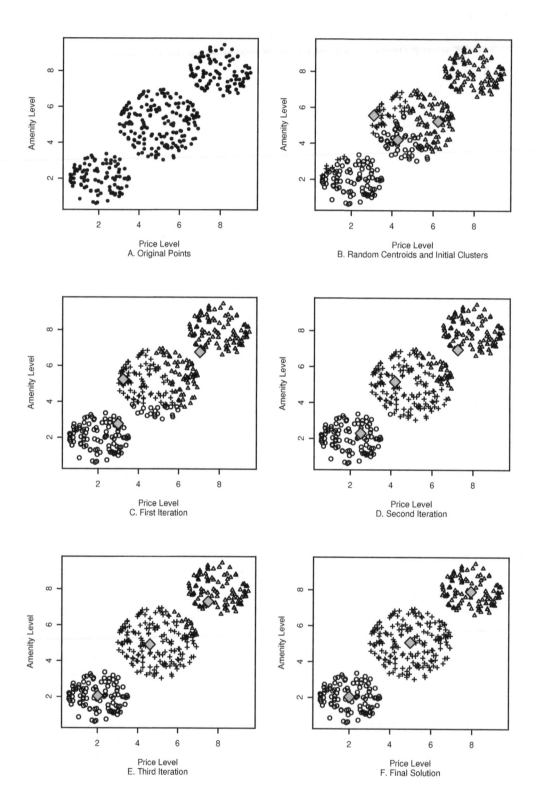

business in which two important service features to customers are the number of amenities they are treated to and the price of the service. Customers usually prefer more amenities and lower prices, but since more amenities cost the company more, they need to know how much customers are willing to pay for their amenities. The question is whether there are different groups of customers with respect to the comparative importance they place on these two aspects of the service, so that an efficient set of offerings for different segments can be developed. Those customers located in the southwest portion of each panel want low prices and relatively few amenities, while those in the northeast portion prefer more amenities and are willing to pay more for them. We can easily see in Panel A of Figure 11.1 that customers fall into three different clusters, and we expect the clustering algorithm to locate these clusters.

Panel B of Figure 11.1 shows the location of three randomly selected points (shown as gray-shaded diamonds) from the set of original points, which act as the initial centroids. The remaining points are given symbols to indicate which cluster they belong to based on which of the three centroids they are closest to, using standard Euclidean distances. A new set of three centroids is calculated by taking the average values of the price and amenity values for points assigned to each cluster. As is shown in Panel C, this causes the centroid that is furthest to the east in Panel B to shift in a northeasterly direction, the centroid in Panel B that is furthest to the south to shift to the southwest, and the centroid furthest to the west in Panel B to move eastward, toward the center. Next, the points are reassigned to clusters based on their proximity to the newly relocated centroids, with the new assignments shown in Panel C. The process of calculating new centroids based on the new cluster assignments of all the points is repeated (with its effect illustrated in Panel D), shifting one centroid further to the southwest, another a bit further to the northeast, and the third closer to the center. The cluster assignments of all points is then reevaluated, and a new set of cluster centroids and subsequent cluster assignments are calculated (Panel E). At some point the movement of cluster centroids across iterations becomes small (in this example, the centroids actually stop moving), and once the movement of the centroids is below a threshold, the iterations stop, and a solution, as shown in Panel E, is returned. The algorithm easily finds the obvious three clusters.

Another commonly used K-Centroids clustering algorithm is known as K-Medians (Leisch, 2006). The difference between the K-Medians and K-Means algorithm is that for K-Medians the centroids are calculated as the median value of variable for a cluster, and the assignment of points to centroids is based on the Manhattan, or city-block, distance (the sum of the absolute values of the difference between the centroid and a point for all variables used to form the clusters). As we have seen before, medians are much less affected by extreme values or "outliers" than means, so that K-Medians often provides more stable clusters (a concept we discuss in Section 11.3) than does K-Means when there are outliers in the data.

The Neural Gas algorithm of Martinetz et al. (1993) is similar to K-Means in that it uses the Euclidean distance between a point and the centroids to assign that point to a particular cluster. However, the method differs from K-Means in how the cluster centroids are calculated in each iteration. Specifically, the location of the centroid for a cluster involves a weighted average of all data points, with the points assigned to the cluster for which the centroid is being constructed receiving the greatest weight, points from the most distant cluster from the focal cluster receiving the lowest weight, and the weights given to points in intermediate clusters decreasing as the distance between the focal cluster and the cluster to which a point is assigned increases. The weights actually follow a soft transfer function used in neural network methods (as illustrated in Figure 8.6) based on distance. We can view K-Means or K-Medians as special cases of this method, where the weights take on the value of one (if a point is assigned to the focal cluster) or zero (if a point is not in the focal cluster), a hard transfer function.[2] The benefit of the Neural Gas algorithm over K-Means is that it is less likely to become "stuck" at a local optima given an initial set of random centroids. The cost is greater computational complexity.

11.2   Cluster Types and the Nature of Customer Segments

The three clusters shown in Figure 11.1 are what Dolnicar and Leisch (2010) describe as "natural" clusters since the three clusters are internally homogeneous and are well separated from one another. The clusters suggest three well-differentiated customer segments: a "high-end" high price and high amenity segment, a "budget" segment looking for a low price in exchange for low amenity levels, and a "mid-market" segment looking for a balance between price and amenity levels. Knowledge of the existence of these three segments has immediate implications for designing service offerings and the marketing communications activities needed to support them.

A survey of 198 marketing managers conducted by Dolnicar and Leisch (2010)

---

[2]It is the framing of the weights used to construct the cluster centroids as transfer functions that leads to the "neural" portion of the algorithm's name. The "gas" portion of the name is based on an analogy made by Martinetz et al. (1993) to the "gas-like" dynamic behavior across iterations of the algorithm.

Figure 11.2: Customer Data on the Interest in Price and Amenity Levels for a Service

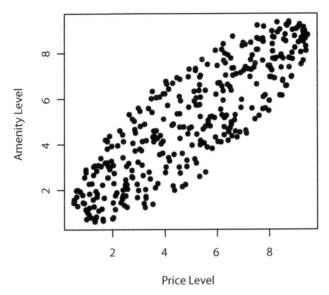

reveals that two-thirds of the managers interviewed believed that cluster analysis studies done to develop customer segments were based on finding a set of natural clusters. Unfortunately, customer data rarely contain a set of natural clusters. Instead, Figure 11.2 provides a hypothetical example of what is more likely to be found. This figure reveals that there are no natural clusters in the data, rather the data are correctly viewed (from a statistical perspective) as a single cluster. Is the statistical view the correct one for the marketing manager to hold? The answer to this question is likely to be "no" since this point of view implies that the manager should either follow a "mass market" approach and develop a single "mid-market" service offering, or develop an offering specific to each customer.[3]

Whether customers fall into a set of natural clusters should actually be of minimal concern to managers; it does, however, complicate the lives of analysts since finding natural clusters, if they exist, greatly simplifies the process of cluster analysis. The real concern for managers should be whether cus-

---

[3]Improvements in information technology and flexible manufacturing have increased the number of different customer segments that firms can tailor targeted product and service offering toward. However, most industries are still a long way from providing product and service offerings that are tailored to a specific customer for any firm that deals with more than a few tens of customers. In the case of B2C firms, it seems unlikely that fully personalized product or service offerings will ever be the case except in a few industries. Put another way, Google will provide increasingly more customized search results for each user, while it seems unlikely that Ford and Toyota will ever be able to provide a car or truck that is truly fully customized for each customer.

Figure 11.3: The K-Means Clusters of the Customer Data for a Service Provider

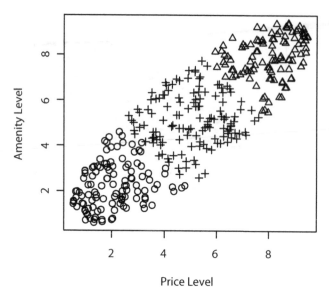

tomers can successfully be grouped into a reasonably small (and thus more manageable) number of segments in order to enhance both firm and customer value. The data in Figure 11.2 help illustrate this point. While there are no natural clusters in the data, it is not without structure, given its southwest-to-northeast orientation. This suggests that placing customers into different groups, and developing multiple service offerings for each of those groups (or choosing a niche by developing one service targeted toward one underserved group) is likely to meet with better success than following a strategy of offering only a single service option that meets the needs of the "average" customer. The issue now is how many segments can customers be systematically divided into, and can data mining tools help in defining segments? To help address these issues, Figure 11.3 illustrates the set of clusters, or segments, defined by applying K-Means to the data (in the next section we will discuss how we determined that a three-cluster solution was appropriate for this data). The three clusters shown in the figure can be described as "budget," "mid-market," and "high-end" segments, with the same underlying interpretation as the three natural clusters shown in Figure 11.2. In this case, even though natural clusters do not exist, customers can still be segmented in a way that enables the creation of more focused product/service offerings and the associated marketing communications programs for those offerings.

In addition to natural clusters, Dolnicar and Leisch (2010) also define what they call "constructed" clusters, and as a subset of constructed clusters, they consider the special case of "reproducible" clusters. Reproducible clusters exist when there is sufficient structure in the data that clusters based on different samples taken from the data result in a consistent, and thus stable, cluster structure (which is the case for the data in Figure 11.3). While the existence of natural clusters in customer data is extremely rare, a reproducible cluster structure can often be found. However, the choice of the clustering method used is a much more critical issue for reproducible clusters compared to natural clusters. Most appropriate clustering methods will be able to uncover natural clusters, however, this is not the case with reproducible clusters. In the case of reproducible clusters, the choice of clustering methods is much more structure-imposing as opposed to being structure-finding. A consequence is that managerial considerations become more obviously important — a good thing. The analyst's task changes from one of finding clusters to one of *creating* clusters that are managerially actionable and meaningful. Practically, it means that the analyst will often select a clustering method to use based on that method's ability to uncover a reproducible cluster structure that results in a managerially useful set of customer segments.

In some cases customer data does not have enough structure to allow for the construction of reproducible clusters. This does not mean that managers should not attempt to create segments. The possibility of creating greater value for both customers and the firm in a way that is managerially tractable still exists. It does mean that cluster analysis tools are likely to be of less value in constructing these segments, but other analytical tools can still help in defining segments (the two- and three-dimensional bi-plot tools presented in the last chapter can be particularly useful in these cases). However, the process of assigning customers to segments becomes more judgmental in nature.

11.3 Methods to Assess Cluster Structure

Based on the discussion of the last section, an important consideration in assessing the results of a K-Centroids clustering is the extent to which a clustering solution is reproducible. Another important consideration, and one that has received more academic research effort, is the extent to which clusters are both internally homogeneous, and separated from one another (the issue that dendrograms address in hierarchical cluster analysis). If customers truly fall into a set of stable clusters, then it should be the case that a set of different random samples of customer data should result in approximately the same set of underlying clusters across the samples except for small differences that are due to both random sampling variability and to randomness induced by the method used to generate the starting set of centroids, via selecting K points at random, in the general K-Centroids algorithm.

In the case of predictive modeling, we addressed the issue of random sampling variability by dividing the database into two or three samples (e.g., the estimation, validation, and holdout samples). For clustering, we really need to look at more than two or three different samples (in the multiple tens or low hundreds). If the database is large enough, we can create enough samples of sufficient size to do this. However, this will frequently not be the case. Instead, following Dolnicar and Leisch (2010), we will make use of a resampling method known as bootstrapping (Efron and Tibshirani, 1994). In bootstrapping, records from the original data set are selected at random, with replacement, to create a bootstrap replicate. This process is repeated a pre-specified number of times to create the desired number of samples,[4] and the data from each sample are then clustered. By convention, the number of records contained in a bootstrap replicate equals the number of records in the original database, however, the convention is fairly arbitrary.

11.3.1 The Adjusted Rand Index to Assess Cluster Structure Reproducibility

A few measures have been proposed to assess the stability of clustering solutions. The one that an extensive simulation study conducted by Milligan and Cooper (1985) found to be the best of these methods was the adjusted Rand index of Hubert and Arabie (1985). The adjusted Rand index provides a measure of similarity between two different clustering solutions, taking a maximum value of one when the two clustering solutions perfectly overlap. The index can be used to determine both the relative and absolute reproducibility of a clustering solution by comparing pairs of solutions, where each pair is based on a different sample of customer data. The greater the overlap between pairs of solutions implies the greater the reproducibility of the cluster structure.

The adjusted Rand index is based on an index first proposed by Rand (1971), and is commonly referred to as the Rand index. The original Rand index takes on values between zero and one, with a value of one indicating a perfect overlap between two solutions, and value of zero indicating that no two points share a common cluster between the two solutions.[5] The adjusted version of the Rand index was created to overcome an important problem with the original.

To provide the intuition behind the adjusted Rand index, it is easiest to start by describing the original Rand index. In addition, a specific example should

---

[4]A similar process was used to generate the cross-validation plots in the tree model.

[5]Obtaining a Rand index value of zero can only occur in situations where there are a large number of clusters relative to the number of points.

Table 11.1: Two Different Cluster Analysis Solutions

Solution V

		V1	V2	Sum
Solution U	U1	8	42	50
	U2	38	12	50
	Sum	46	54	100

Table 11.2: Calculated Unique Pairs of Points

Solution V

		V1	V2	Row
Solution U	U1	$\binom{8}{2} = 28$	$\binom{42}{2} = 861$	$\binom{50}{2} = 1225$
	U2	$\binom{38}{2} = 703$	$\binom{12}{2} = 66$	$\binom{50}{2} = 1225$
	Column	$\binom{46}{2} = 1035$	$\binom{54}{2} = 1431$	$\binom{100}{2} = 4950$

help make the discussion of the original index and the adjusted version clearer, which is presented (in contingency table form) in Table 11.1. The example consists of two cluster solutions (labeled U and V), each involving two clusters for a data set with 100 points. The values in the interior cells of the table give the number of points that are in cluster i for solution U and cluster j for solution V. An examination of the table reveals that cluster 1 of solution U roughly corresponds to cluster 2 of solution V, while cluster 2 of solution U and cluster 1 of solution V correspond to each other. The differences between solution U and V could be due to the use of different clustering methods applied to the same data, or (as in what follows) based on clustering using the same algorithm, but with different samples drawn from the original data.

Both the original and adjusted Rand index values are based on the number of pairs of points that are clustered in the same (and different) way across the two solutions. The number of unique pairs in a set of n points is calculated using the binomial coefficient (Wikipedia, 2011), and the results of those calculations for the cells, row sums, column sums, and the entire data set for solutions U and V are given in Table 11.2.

To calculate the Rand index, we need to use the values in Table 11.2 to calculate four additional values. The first is the total number of point pairs that are in cluster i of solution U and cluster j of solution V across all the possible combinations of i and j (i.e., over the interior cells of the contingency table). This count of the common pairs is labeled as a, and can be stated mathematically as

$$a = \sum_{i,j} \binom{n_{ij}}{2},$$

where n_{ij} is the number of points that are assigned to both cluster i by solution U and cluster j by solution V. For the example, taking advantage of the values presented in Table 11.2, the value of a is $28 + 861 + 703 + 66 = 1658$.

The second value that needs to be calculated for the Rand index is the number of point pairs that are placed in the same cluster for solution U, but in different clusters for solution V. This value (which is labeled as b) can be calculated by taking the *difference* between the total number of point pairs that were placed in the same cluster by solution U and the total number of point pairs that were placed in the sample cluster for both solutions U and V (which correspond to a). Mathematically, this can be written as

$$b = \sum_{i} \binom{n_{i.}}{2} - a,$$

where $n_{i.}$ is the number of points assigned to cluster i by solution U. In our example, this corresponds to $(1225 + 1225) - 1658 = 792$.

A nearly identical calculation is done to determine the third needed value, the number of point pairs that are placed in the same cluster of solution V, but not in the same cluster of solution U (which is labeled as c), or

$$c = \sum_{j} \binom{n_{.j}}{2} - a,$$

where $n_{.j}$ is the number of points assigned to cluster j in solution V. The value of c in our running example is $(1035 + 1431) - 1658 = 808$.

The final value (labeled as d) that is needed to calculate the Rand index is the number of pairs of points that are in different clusters in both solutions. This value can be calculated as the difference between the total number of possible point pairs in the data set and the other three calculated values we have just described (a, b, and c), or

$$d = \binom{n}{2} - a - b - c,$$

where n is the total number of points in the data set. For our running example, this corresponds to $4950 - 1658 - 792 - 808 = 1692$.

The Rand index (RI) is given by

$$RI = \frac{a + d}{a + b + c + d} = \frac{a + d}{\binom{n}{2}},$$

which for our example corresponds to $(1658 + 1692)/4950 = 0.676$. If we altered the example slightly, replacing the interior cells of the contingency table in Table 11.1 with the values $n_{11} = 0$, $n_{12} = 50$, $n_{21} = 50$, and $n_{22} = 0$, then the Rand index equals one since the values of b and c both equal zero. One thing to notice is that the value of a plays a prominent role in the Rand index, and this is taken advantage of in the creation of the adjusted Rand index.

A major concern with the original Rand index is that the expected value of the index for two randomly created clustering solutions does not take a constant value (such as zero). To address this issue, Hubert and Arabie (1985) proposed an adjusted Rand index. This index starts with the general form of an adjusted index and is based on the counts of common pairs, a, with a constant expected value, given by

$$Adjusted\ Index = \frac{a - Expected[a]}{Maximum[a] - Expected[a]},$$

where $Expected[a]$ is the expected value given the underlying distribution of possible values of a. Hubert and Arabie (1985) assume the underlying distribution of a is generalized hypergeometric, allowing them to show that

$$Expected[a] = E\left[\sum_{i,j} \binom{n_{ij}}{2}\right] = \frac{\sum_i \binom{n_{i.}}{2} \sum_j \binom{n_{.j}}{2}}{\binom{n}{2}}.$$

Finally, the maximum value is given by

$$Maximum[a] = \frac{1}{2}\left[\sum_i \binom{n_{i.}}{2} + \sum_j \binom{n_{.j}}{2}\right].$$

Substituting these terms into the general form of an adjusted index allows the adjusted Rand index (ARI) to be written as

$$ARI = \frac{\sum_{i,j} \binom{n_{ij}}{2} - \left[\sum_i \binom{n_{i.}}{2} \sum_j \binom{n_{.j}}{2}\right]/\binom{n}{2}}{\frac{1}{2}\left[\sum_i \binom{n_{i.}}{2} + \sum_j \binom{n_{.j}}{2}\right] - \left[\sum_i \binom{n_{i.}}{2} \sum_j \binom{n_{.j}}{2}\right]/\binom{n}{2}}.$$

For our example, we already know that the value of a is 1648, while

$$Expected[a] = \frac{(1225 + 1225)(1035 + 1431)}{4950} = 1220.545$$

and

$$Maximum[a] = \frac{1}{2}(1225 + 1225 + 1035 + 1431) = 2458.$$

Entering these values into the *ARI* formula results in an adjusted Rand index value of $(1658 - 1220.545)/(2458 - 1220.545) = 0.354$ for the example.

The RcmdrPlugin.BCA package implements an adjusted Rand index assessment for different numbers of clusters in the solution for a particular K-Centroids clustering algorithm using a bootstrapping approach that is implemented in the flexclust package (Dolnicar and Leisch, 2010; Leisch, 2006). The user specifies the number of paired-comparisons of clustering solutions to be used and the range of K values (for example, clustering solutions with between two and six clusters) to be assessed. For each paired-comparison two bootstrap samples are drawn, clustering solutions for each value of K for both bootstrap samples are determined, the cluster assignments for the full data set under each solution are determined, and the adjusted Rand index values are calculated for each value of K. This process is repeated for each of the user-selected number of paired comparisons, so if the user specified 100 paired comparisons should be made, it is done 100 times. The end result is a boxplot of the adjusted Rand index values for each value of K in the range of cluster solutions to be assessed.

To help fix ideas, Figure 11.4 contains the adjusted Rand index boxplots for the data shown in Figure 11.3 (the elliptically shaped customer data for a service). In this example, solutions based on the K-Means algorithm using between two and six clusters are examined, and 100 paired comparisons of the clustering solutions are made. We are interested in using the boxplots to determine the number of clusters that consistently result in relatively high values of the adjusted Rand index. An examination of Figure 11.4 indicates that solutions with either two- or three-clusters have relatively high values of the adjusted Rand index, with the adjusted Rand index sharply falling with more than three clusters. This indicates that the two- and three-cluster solutions are more reproducible than solutions involving a larger number of clusters. As a result, selecting either the two- or three-cluster solution is a reasonable choice on the basis of cluster reproducibility. However, for managerial purposes, it is likely to make sense to select the three-cluster solution since the clusters make logical sense (the solution leads to the intuitive customer segments of "high-end," "mid-market," and "budget"), is likely to result in service offerings that are more closely aligned with customer preferences than would be the case if

Figure 11.4: Boxplot of the Adjusted Rand Index for the Elliptical Customer Data

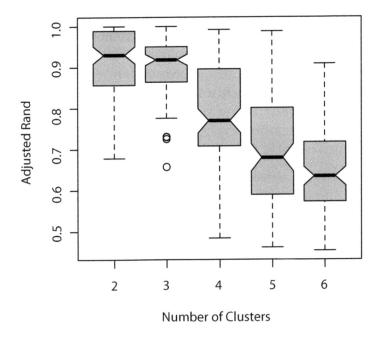

the market was divided into only two segments, and still simplifies the world enough to be managerially tractable.

As one may have already guessed, with a large range in the number of clusters, a large data set, and a large number of paired comparisons, the assessment can take a noticeable amount of computer time to complete (on the order of what a step-wise regression model can take). While reducing the number of comparisons greatly reduces the amount of computer time needed to do an assessment, it also reduces the reliability of that assessment. The flexclust package takes 100 as the default value of the number of paired comparisons (resulting in the creation of 200 bootstrap replicate samples), and we encourage users to select the default value. A more promising avenue to reduce the amount of computer time needed is to decrease the range of K values to consider in a run, starting with between two and six clusters, and then extend the range out (say from six to ten) if the adjusted Rand index values continue to improve as the number of clusters is increased.[6]

---

[6]Users on Mac OS X, Linux, and other POSIX compliant operating systems (Windows is *not* POSIX compliant) with quad core processors in their computers can get a substantial improvement in speed by installing the multicore package (Urbanek, 2011), which enables the simultaneous use of multiple processor cores for the analysis. The multicore package can be installed by going the R Console command line and entering the command: install.packages("multicore"), and then pressing Enter.

11.3.2 The Calinski-Harabasz Index to Assess within Cluster Homogeneity and between Cluster Separation

A large number of different measures have been proposed to assess the extent to which clusters are relatively internally homogeneous and separated from one another. One of the most commonly used measures to assess this is the Calinski–Harabasz index (also known as the "pseudo F-statistic" and the "variance ratio criteria") which was developed by Calinski and Harabasz (1974), and was shown to be the best single measure at finding the actual number of clusters in the simulation experiments of Milligan and Cooper (1985), with similar findings recently reported by Vendramin et al. (2009).

The calculation of the Calinski–Harabasz index is based on comparing the weighted ratio of the between cluster sum of squares (the measure of cluster separation) and the within cluster sum of squares (the measure of how tightly packed the points are within a cluster). Ideally, the clusters should be well separated, so the between cluster sum of squares value should be large, but points within a cluster should be as close as possible to one another, resulting in smaller values of the within cluster sum of squares measure. Since the Calinski–Harabasz index is a ratio, with the between cluster sum of squares in the numerator and the within cluster sum of squares in the denominator, cluster solutions with larger values of the index correspond to "better" solutions than cluster solutions with smaller values. In practice, we will be interested in finding the number of clusters that has the largest value of the index holding the cluster algorithm used constant.[7]

It turns out that we saw the within cluster sum of squares measure (which we label WCSS) used in the Calinski–Harabasz index when the clustering algorithm used is K-Means or Neural Gas in Section 10.2, but by a different name; there it was called the error sum of squares (or ESS).[8] The between cluster sum of squares (which we label BCSS) is calculated using the relationship that

$$TSS = WCSS + BCSS,$$

where TSS is the total sum of squares. In the case of K-Means and Neural Gas, and using the notation of Section 10.2, the total sum of squares can also

---

[7]The adjusted Rand index can actually be used for making comparisons across different algorithms, which is not the case for the Calinski–Harabasz index since the adjusted Rand index is distance metric independent.

[8]For K-Means the within cluster sum of squares and error sum of squares are one and the same. In the case of Neural Gas, the cluster centroids (based on the weighted average of all points) is used instead of the cluster means. For consistency purposes, in the case of K-Medians, the cluster medians are used as the cluster centers and the sum is done based on the use of the Manhattan distance measure.

be calculated as

$$\text{TSS} = \sum_i D^2_{c,x_i},$$

where c is the centroid of the the entire data set (the mean vector for the K-Means and Neural Gas algorithms and the median for the K-Medians algorithm), and D_{c,x_i} is the distance between the data set centroid and point x_i. The value of BCSS is found by taking the difference between TSS and WCSS. The normalization used to adjust the WCSS value is $n - K$ (the difference between the total number of points in the data and the number of clusters in the solution), and BCSS is normalized by $K - 1$, resulting in

$$\text{Calinski-Harabasz Index} = \frac{\text{BCSS}/K - 1}{\text{WCSS}/n - K}.$$

As with the adjusted Rand index, the RcmdrPlugin.BCA package makes use of bootstrap replicates of Calinski–Harabasz indices to address randomness due to random sampling error and the use of random initial centroids. The software makes use of the bootstrap samples created for the adjusted Rand index assessment. Unlike the adjusted Rand index, the Calinski–Harabasz index is calculated for each solution rather than for a paired comparison. As a result, if 100 comparisons are used for the adjusted Rand index assessment, 200 Calinski–Harabasz index values are calculated.

The Calinski–Harabasz indices are presented as boxplots for each value of K analyzed. For the example data contained in Figure 11.3 (the service provider's customer data) the indices are shown in Figure 11.5, based on 200 bootstrap replicates (the same 200 used to create the 100 adjusted Rand index comparisons) using the K-Means algorithm with two to six clusters. An examination of Figure 11.5 strongly indicates that the three-cluster solution is the "best" solution with respect to its relative separation between clusters and homogeneity within clusters based on the Calinsk–Harabasz index. When these results are combined with those of the adjusted Rand index analysis and the managerial argument, a three-segment solution appears best and managerially useful, despite the absence of natural clusters in the customer data.

11.4   K-Centroids Clustering Tutorial

In this tutorial you will see how to use R Commander to perform a K-Centroids cluster analysis based on the use of the K-Means algorithm, and make an assessment of the number of clusters to use in the final solution. We will compare the K-Means and Ward's Method clusters, and determine which provides

Figure 11.5: Boxplot of the Calinski–Harabasz Index for the Elliptical Customer Data

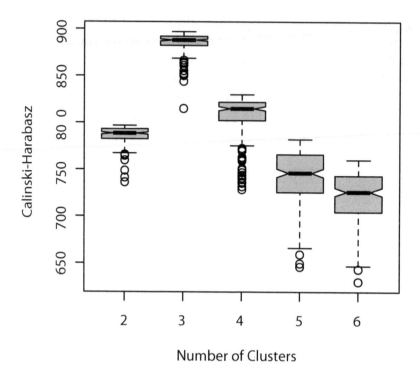

a "better" solution based on interpreting the clusters, and how cohesive the clusters are for each solution.

1.

If you are continuing from the Ward's Method tutorial without quitting from R Commander you should have the Athletics data set with new variables available and be ready to go. If you have started R Commander from the Ward's workspace file, you will have to activate the Athletic data set by clicking on the Data Set button (now labeled <No active data set>) and selecting it. You should also redo the Wards method four-cluster solution and add the cluster assignments to the data set.

2.

We chose four clusters based on the dendrogram of the Ward's method solution, so we will begin by assessing the cluster structure for different numbers of K-Means clusters around four using both the adjusted Rand index to assess cluster reproducibility and the Calinski–Harabasz index to assess relative

Figure 11.6: The K-Centroids Clustering Diagnostics Dialog Box

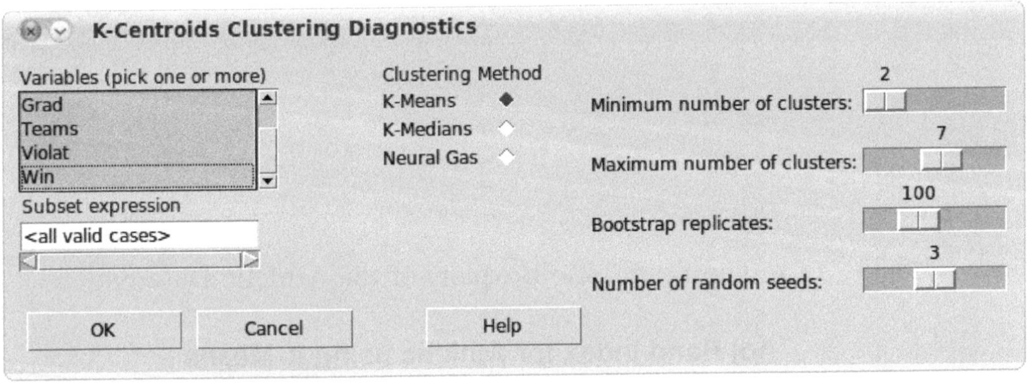

within-cluster homogeneity and across-cluster separation. To do this assessment, use the pull-down menu option **Group → Records → k-centroids clustering diagnostics...**, which will bring up the dialog box shown in Figure 11.6.

3.

In the dialog box for the "Variables (pick one or more)" **select the seven importance weights** for intercollegiate athletic program success. For the clustering method, select K-Means (at the end of tutorial we will consider the other two methods). The Ward's method dendrogram indicated there were four reasonably defined clusters in the data. For K-Means we are going to bracket that solution for a range from two to seven clusters. Thus, **select 2 as the minimum** number of clusters and **7 as the maximum** number of clusters using the sliders. The other two options "Boostrap replicates:" and "Number of random seeds:" control aspects of the assessment in R. Issues surrounding the choice of the number of bootstrap replicates are discussed at the end of Section 11.3.1, while the number of random seeds controls the number of times a clustering algorithm is run for each bootstrap replicate, with the final solution for the replicate being the best of those runs. The default settings work well in our case. After **pressing OK**, and after a bit of time, the boxplots shown in Figure 11.7 should appear.

4.

The plot in Figure 11.7 shows a clear maximum of both the adjusted Rand index and the Calinski–Harabasz index at four clusters. This matches the Ward's method dendrogram results of the last lab, giving us further confidence that the four-cluster solution should be selected.

Figure 11.7: The Diagnostic Boxplots of the Athletic Data Set

Adj Rand Index for Athletic using K-Means

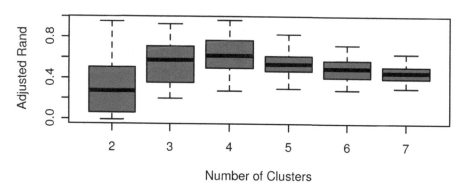

C-H Index for Athletic using K-Means

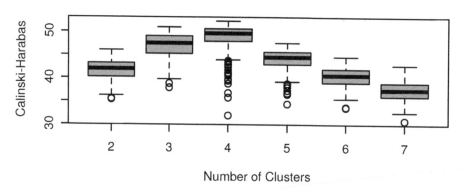

Figure 11.8: The K-Centroids Clustering Dialog Box

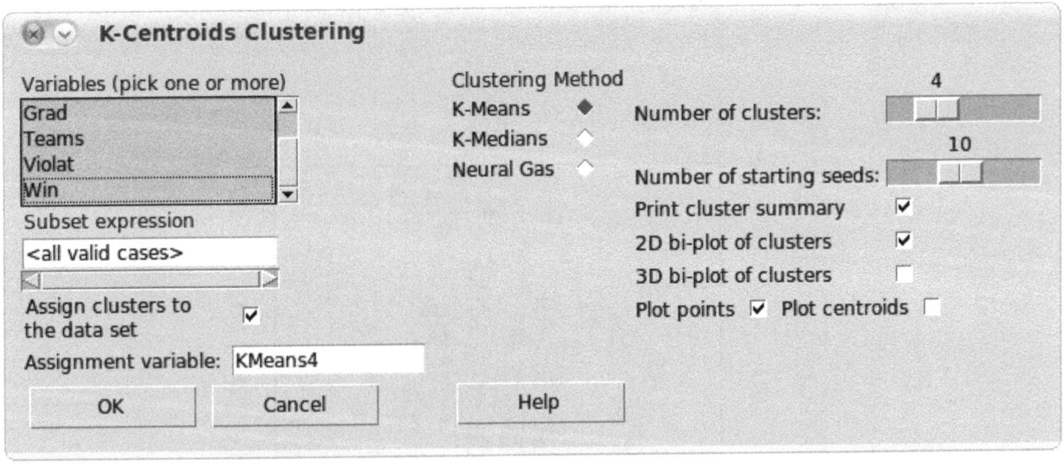

5.

Use the pull-down menu option **Group → Records → k-centroids cluster analysis...**, which will cause the dialog shown in Figure 11.8 to appear. Again **select the seven importance** weights in the "Variables (pick one or more)" option, select K-Means as the clustering method, and use the slider to **select 4** for the "Number of clusters:" Leave "Number of starting seeds:" at its default value. To compare the K-Means and Ward's method clustering solutions, **check the "Assign clusters to the data set"** option and **call the new cluster assignment variable KMeans4.** Maintain the defaults of printing a cluster summary and creating a two-dimensional bi-plot of the solution. After **pressing OK,** the bi-plot (Figure 11.9) and cluster statistical summary will be printed (Figure 11.10), and the cluster assignment variable, KMeans4, will be added to the data set.

6.

An examination of the bi-plot in Figure 11.9 indicates that the K-Means clustering solution is very similar to that obtained by Ward's method, with one cluster for those who view win/loss records as most important, one that views intercollegiate athletic department financial performance as most important, one that views NCAA rule violations as most important, and one that views student athlete graduation rates as most important. The numbers corresponding to these clusters is different for the two solutions (and your K-Means clusters could be numbered differently from the ones shown in Figure 11.9), but the cluster numbering is arbitrary and not meaningful. A close examination of the two bi-plots (Figure 10.14 and Figure 11.9) suggests that the K-Means clusters are a little bit more cohesive than the Ward's method clusters, but

Figure 11.9: The Bi-Plot of the Four-Cluster K-Means Solution

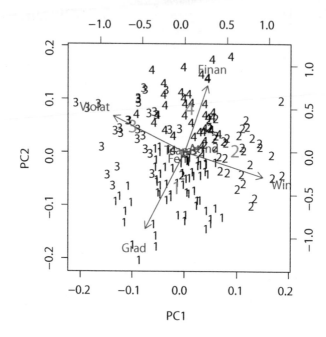

Figure 11.10: The Statistical Summary of the Four-Cluster K-Means Solution

```
> summary(.cluster)
kcca object of family 'kmeans'

call:
stepFlexclust(x = model.matrix(~-1 + Attnd + Fem + Finan + Grad +
    Teams + Violat + Win, Athletic), k = 4L, nrep = 10, FUN = kcca,
    family = kccaFamily("kmeans"))

cluster info:
  size   av_dist   max_dist  separation
1   66 0.1562185 0.3672764   0.1520345
2   35 0.1438229 0.2567440   0.1615703
3   29 0.1422995 0.2555475   0.1845425
4   38 0.1644309 0.3827415   0.1665350

convergence after 16 iterations
sum of within cluster distances: 25.71928

> print(.cluster@centers) # Cluster Centroids
          Attnd        Fem      Finan       Grad      Teams     Violat        Win
[1,] 0.08833036 0.08323994 0.1670436 0.2977241 0.06422680 0.1280949 0.1713403
[2,] 0.08775407 0.07160773 0.2325328 0.1211496 0.05324482 0.1018854 0.3318256
[3,] 0.06326663 0.06593792 0.1828255 0.1704187 0.05672456 0.3577898 0.1030368
[4,] 0.12548015 0.07281624 0.3362693 0.1352190 0.06595774 0.1480240 0.1128954
```

Figure 11.11: The Overlap between the K-Means and Ward's Method Solutions

Kmeans4	Wards4 (1)	2	3	4
1	5	61	0	0
(2)	31	5	1	0
3	2	1	22	4
4	15	3	0	18

only very slightly. A comparison of the summaries for the two-cluster solutions (Figure 10.10 and Figure 11.10) also indicates a high degree of similarity between the two sets of clusters based on their centroids. Overall, we have a slight preference for the K-Means solution, largely because the cluster most concerned with financial performance and student athlete graduation rates seems to have slightly sharper boundaries in this solution.

7.

The last way to compare the two-cluster solutions is to do a two-way contingency table to compare the cluster membership variables. Use the pull-down menu option **Explore and test → Contingency tables → Two-way table** to create the table. In this instance the chi-squared test of independence is not useful (we are not interested in generalizing a relationship between variables to a larger population!), so you may want **to uncheck this option**. Your contingency table should look **similar** (but probably not identical) to the one shown in Figure 11.11. Use the two bi-plots to identify the correspondence between the K-Means cluster numbers and the Ward's numbers. For example, in this particular run, the cluster associated with the WIN Principle Component is Ward's **1** and K-Means **2** (identified by circles). Ward's method places 53 individuals in this cluster, and K-Means places 37 in it (the total for the row and column in rectangles). The overlap between the two is 31 individuals, who are in both the Ward's 1 and K-Means 3 clusters. In other words, the two methods classify 31 individuals in the same WIN cluster.

8.

Exercise: Find the correspondence for the other three clusters and confirm that in this case 36 of 168 of the respondents are placed in different clusters by the two methods, and 132 in the same cluster, a high degree of overlap. Express this overalap as the percentage of the 168 individuals who are assigned to the same clusters for the two methods. The fact that these two methods have such high agreement inspires confidence that that a four-segment solution based on attitudes toward athletic programs is a good way to view this "market" should you want to take some action such as designing a communications program.

9.

Exercise: Repeat steps 2, 3, 5, and 7, but select K-Medians as the clustering method, and save the cluster assignments to a suitably named new variable. Create a contingency table comparing the K-Medians results to the K-Means results. Repeat the process a final time, but this time selecting Neural Gas as the clustering method. Based on this analysis, which clustering solution would you select as best and why?

10.

Exit R and R Commander. You won't need any of the results, but it is probably a good idea to save your workspace for future reference.

Bibliography

Amemiya, Takeshi. 1985. *Advanced Econometrics.* Harvard University Press.

Barry, Michael J. A., and Linoff, Gordon. 1997. *Data Mining Techniques of Marketing, Sales, and Customer Relationship Management.* John Wiley & Sons.

Breiman, Leo, Friedman, Jerome H., Olshen, Richard A., and Stone, Charles J. 1984. *Classification and Regression Trees.* Wadsworth.

Bulmer, Micheal. 2003. *Francis Galton: Pioneer of Heredity and Biometry.* Johns Hopkins University Press.

Calinski, T., & Harabasz, J. 1974. A Dendrite Method for Cluster Analysis. *Communications in Statistics,* **3**(1), 1–27.

Chapman, Peter, Clinton, Julian, Kerber, Randy, Khabaza, Thomas, Reinartz, Thomas, Shearer, Colin, and Wirth, Rüdiger. 2000. *CRISP-DM 1.0: Step-By-Step Data Mining Guide.* SPSS.

Cleveland, William S., and Devlin, Susan J. 1988. Locally Weighted Regression: An Approach to Regression Analysis by Local Fitting. *Journal of the American Statistical Association,* **83**, 596–610.

Das, Mamuni. 2003. *How Verizon Cut Customer Churn by 0.5 Per Cent.* Web. accessed on October 11, 2006.

Davenport, Thomas H. 2006. Competing on Analytics. *Harvard Business Review,* January, 99–107.

Dolnicar, Sara, and Leisch, Friedrich. 2010. Evaluation of Structure and Reproducibility of Cluster Solution Using the Bootstrap. *Marketing Letters,* **21**(1), 83–101.

Efron, Bradley, and Tibshirani, Robert. 1994. *An Introduction to the Bootstrap.* Chapman & Hall/CRC.

Fellows, Ian. 2011. *Deducer: Deducer.* R package version 0.5-0.

Fox, John. 2005. The R Commander: A Basic-Statistics Graphical User Interface to R. *Journal of Statistical Software*, **14**(9), 1–42.

Graettinger, Tim. 1999. *Digging Up Dollars with Data Mining: An Executive's Guide*. The Data Administration Newsletter-TDAN.com, September.

Guha, Sudipto, Rastogi, Rajeev, and Ship, Kyuseok. 2000. ROCK: A Robust Clustering Algorithm for Categorical Attributes. *Information Systems*, **25**(5), 345–366.

Hebb, D. O. 2002. *The Organization of Behavior: A Neuropsychological Theory*. Lawrence Erlbaum Associates.

Hubert, Lawrence, and Arabie, Phipps. 1985. Comparing Partitions. *Journal of Classification*, **2**(2), 193–218.

Judge, George G., Hill, R. Carter, Griffiths, William, Lütkepohl, Helmut, and Lee, Tsoung-Chao. 1982. *Introduction to the Theory and Practice of Econometrics*. John Wiley & Sons.

Kelly, Jack. 2003. Data and Danger: Let the Government Connect the Dots. *The Pittsburgh Post-Gazette*, October 10.

Krivda, Cheryl D. 1996. Unearthing Underground Data. *LAN Magazine*, **11**(5), 12–18.

Leisch, Friedrich. 2006. A Toolbox for K-Centroids Cluster Analysis. *Computational Statistics and Data Analysis*, **51**(2), 526–544.

MacQueen, J. 1967. Some Methods for Classification and Analysis of Multivariate Observations. Pages 281–297 of: Cam, L. M. L., and Neyman, J. (eds.), *Proceedings of the Fifth Berkeley Symposium on Mathematical Statistics and Probability*.

Mardia, K. V., Kent, J. T., and Bibby, J. M. 1979. *Multivariate Analysis*. Academic Press.

Martinetz, Thomas M., Berkovich, Stanislav G., and Schulten, Klaus J. 1993. "Neural-Gas" Network for Vector Quantization and its Application to Time-Series Prediction. *IEEE Transactions on Neural Networks*, **4**(4), 558–569.

McCullagh, P., & Nelder, J. A. 1989. *Generalized Linear Models*. 2 edn. Chapman and Hall.

McFadden, D. 1974. The Measurement of Urban Travel Demand. *Journal of Public Economics*, **3**(3), 303–328.

Milligan, Glenn W., and Cooper, Martha C. 1985. An Examination of Procedures to Determine the Number of Clusters in a Data Set. *Psychometrika*, **50**(2), 159–179.

Nelder, J. A., and Wedderburn, R. W. M. 1972. Generalized Linear Models. *Journal of the Royal Statistical Society, Series A*, **135**(3), 370–384.

Peppers, Don, and Rogers, Martha. 1993. *The One-To-One Future: Building Relationships One Customer at a Time*. Currency-Doubleday.

R Development Core Team. 2011. *R: A Language and Environment for Statistical Computing*. R Foundation for Statistical Computing, Vienna, Austria. ISBN 3-900051-07-0.

Rand, W. M. 1971. Objective Criteria for the Evaluation of Clustering Methods. *Journal of the American Statistical Association*, **66**(336), 846–850.

Rexer, Karl, Allen, Heather, and Gearan, Paul. 2010. 2010 Data Miner Survey Summary. Presented at Predictive Analytics World, October.

Urbanek, Simon. 2011. *multicore: Parallel Processing of R Code on Machines with Multiple Cores or CPUs*. R package version 0.1-7.

Venables, W. N., and Ripley, B. D. 2002. *Modern Applied Statistics with S*. Springer.

Vendramin, Lucas, Campell, Ricardo J. G. B., and Hruschka, Eduardo R. 2009. On the Comparison of Relative Clustering Validity Criteria. In: *Proceedings of the Ninth SIAM International Conference on Data Mining*, vol. 9. SIAM.

Verzani, John. 2011. *pmg: Poor Man's GUI*. R package version 0.9-42.

Vesset, Dan, and McDonough, Brian. 2006. *Worldwide Business Intelligence Tools 2005*. Tech. rept. IDC.

Ward, Joe H. 1963. Hierarchical Grouping to Optimize an Objective Function. *Journal of the American Statistical Association*, **58**(March), 236–244.

Wikipedia. 2011. *Binomial Coefficient — Wikipedia, The Free Encyclopedia*. http://en.wikipedia.org/w/index.php?oldid=459460281.

Williams, Graham J. 2009. Rattle: A Data Mining GUI for R. *The R Journal*, **1**(2), 45–55.

Index